Artificial Neural Networks for Engineering Applications

Artificial Neural Networks for Engineering Applications

Edited by

Alma Y. Alanis
Nancy Arana-Daniel
Carlos López-Franco

University of Guadalajara, CUCEI
Computer Science Department
Guadalajara, Mexico

ELSEVIER

ELSEVIER

3251 Riverport Lane
St. Louis, Missouri 63043

Artificial Neural Networks for Engineering Applications   ISBN: 978-0-12-818247-5

Publisher: Mara Conner
Acquisition Editor: Chris Katsaropoulos
Editorial Project Manager: Ana Claudia Garcia
Production Project Manager: Nirmala Arumugam
Designer: Greg Harris

List of Contributors

A. Bassam, PhD
Universidad Autónoma de Yucatán
Facultad de Ingeniería
Mérida, Yucatán, México

Victor H. Benitez, PhD
Universidad de Sonora
Department of Industrial Engineering
Mechatronics Area
Hermosillo, México

C. Bhowmick, MS
Missouri University of Science and Technology
Department of Electrical Engineering
Rolla, MO, United States

Daniel Carrillo, DSc
Departamento de Control Automático
CINVESTAV-IPN
Mexico City, Mexico

Hugo Coss y León, MSc
Water and Energy Department
CUTONALA
Universidad de Guadalajara
Tonalá, Jalisco, México

E. Cruz May, MEng
Universidad Autónoma de Yucatán
Facultad de Ingeniería
Mérida, Yucatán, México

Javier Gomez-Avila, PhD
University of Guadalajara
Laboratory of Intelligent Systems
Guadalajara, Mexico

Rafael González, MEng
Water and Energy Department
CUTONALA
Universidad de Guadalajara
Tonalá, Jalisco, México

Kelly J. Gurubel, DSc
Water and Energy Department
CUTONALA
Universidad de Guadalajara
Tonalá, Jalisco, México

Gustavo Hernandez-Mejia, PhD
Frankfurt Institute for Advanced Studies
Frankfurt am Main, Germany

Esteban A. Hernandez-Vargas, PhD
Frankfurt Institute for Advanced Studies
Frankfurt am Main, Germany

S. Jagannathan, PhD
Missouri University of Science and Technology
Department of Electrical Engineering
Rolla, MO, United States

Francisco Jurado, DSc
Tecnológico Nacional de México/I.T. La Laguna
Div. de Est. de Posg. e Invest.
Torreón, Coahuila de Zaragoza, México

Sergio Lopez, MSc
Tecnológico Nacional de México/I.T. La Laguna
Div. de Est. de Posg. e Invest.
Torreón, Coahuila de Zaragoza, México

Gehová López-González, PhD
University of Guadalajara
Laboratory of Intelligent Systems
Guadalajara, Mexico

O. May Tzuc, MEng
Universidad Autónoma de Yucatán
Facultad de Ingeniería
Mérida, Yucatán, México

Haifeng Niu, PhD
Missouri University of Science and Technology
Department of Electrical Engineering
Rolla, MO, United States

Fernando Ornelas-Tellez, PhD
Universidad Michoacana de San Nicolas de Hidalgo
Faculty of Electrical Engineering
Morelia, Mexico

Roxana Recio, MSc
Water and Energy Department
CUTONALA
Universidad de Guadalajara
Tonalá, Jalisco, México

L.J. Ricalde, PhD
Universidad Autónoma de Yucatán
Facultad de Ingeniería
Mérida, Yucatán, México

J. Jesus Rico-Melgoza, PhD
Universidad Michoacana de San Nicolas de Hidalgo
Faculty of Electrical Engineering
Morelia, Mexico

Edgar N. Sanchez, DSc
Automatic Control Department
CINVESTAV Guadalajara
Zapopan, Jalisco, México

Angel E. Villafuerte, PhD
Universidad Michoacana de San Nicolas de Hidalgo
Faculty of Electrical Engineering
Morelia, Mexico

Carlos Villaseñor, PhD
University of Guadalajara
Laboratory of Intelligent Systems
Guadalajara, Mexico

Wen Yu, DSc
Departamento de Control Automático
CINVESTAV-IPN
Mexico City, Mexico

Febe J. Zavala-Mendoza, MSc
Universidad Michoacana de San Nicolas de Hidalgo
Faculty of Electrical Engineering
Morelia, Mexico

Nancy Arana-Daniel dedicates this book to her husband, Angel,
and her children, Ana, Sara, and Angel, as well as her parents, Maria and Trinidad,
and her brothers and sisters, Rodolfo, Claudia, Nora, Carlos, Ernesto, Gerardo, and Paola.

Alma Y. Alanis dedicates this book to her husband, Gilberto, her mother, Yolanda,
and her children, Alma Sofia and Daniela Monserrat.

Carlos López-Franco dedicates this book to his wife, Paty,
and his children, Carlos Alejandro, Fernando Yhael, and Íker Mateo.

About the Editors

Alma Y. Alanis was born in Durango, Durango, Mexico, in 1980. She received her BSc degree from the Instituto Tecnológico de Durango (ITD), Durango Campus, Durango, Durango, in 2002, and her MSc and PhD degrees in electrical engineering from the Advanced Studies and Research Center of the National Polytechnic Institute (CINVESTAV-IPN), Guadalajara Campus, Mexico, in 2004 and 2007, respectively. Since 2008 she has worked for the University of Guadalajara, where she is currently a Chair Professor in the Department of Computer Science. She is also a member of the Mexican National Research System (SNI-2) and a member of the Mexican Academy of Sciences. She has published papers in recognized international journals and conference proceedings, besides four international books. She is a Senior Member of the IEEE and Subject and Associated Editor of the Journal of Franklin Institute (Elsevier) and Intelligent Automation and Soft Computing (Taylor and Francis). Moreover she is currently serving on a number of IEEE and IFAC Conference Organizing Committees. In 2013, she received the grant for women in science by L'Oreal-UNESCO-AMC-CONACYT-CONALMEX. In 2015, she received the Research Award Marcos Moshinsky. Since 2008 she is a member of the Accredited Assessors record RCEA-CONACYT, evaluating a wide range of national research projects. Besides, she has belonged to important project evaluation committees of national and international research projects. Her research interest focuses on neural control, backstepping control, block control, and their applications to electrical machines, power systems, and robotics.

Nancy Arana-Daniel received her BSc degree from the University of Guadalajara in 2000 and her MSc and PhD degrees in electric engineering with a specialization in computer science from the Research Center of the National Polytechnic Institute and Advanced Studies, CINVESTAV, in 2003 and 2007, respectively. She is currently a research fellow at the University of Guadalajara, Mexico, in the Department of Computer Science, where she is working at the Laboratory of Intelligent Systems and the Research Center for Control Systems and Artificial Intelligence. She is an IEEE Senior Member and a member of the National System of Researchers (SNI-1). She has published several papers in international journals and conference proceedings and she has been technical manager of several projects that have been funded by the National Council of Science and Technology (CONACYT). Also, she has collaborated in an international project funded by OPTREAT. She is Associated Editor of the Journal of Franklin Institute (Elsevier). Her research interests focus on applications of geometric algebra, geometric computing, machine learning, bio-inspired optimization, pattern recognition, and robot navigation.

Carlos López-Franco received his PhD degree in Computer Science in 2007 from the Center of Research and Advanced Studies, CINVESTAV, Mexico. He is currently a professor at the Computer Science Department of the University of Guadalajara, Mexico, and a member of the Intelligent Systems group. He is IEEE Senior Member and a member of the National System of Researchers (SNI-1). His research interests include geometric algebra, computer vision, robotics, and intelligent systems.

Preface

This book presents efficient, robust, and adaptable neural network–based methodologies to solve important engineering problems. The proposals range from the control of wastewater treatment, quadrotor UAVs, biomechatronic systems, the modeling of dynamical systems, noise filtering and image improvement, the identification of models of disease progression, predicting biofuel production, aiding the operation of photovoltaic systems, and detecting attacks for cyber-physical systems to forecasting time series. The book aims to introduce the artificial neural network theory as well as the design of solutions that use it to deal with real and complex problems.

Chapter 1 presents the modeling of a wastewater treatment plant and its subsystems using a hierarchical dynamic neural network that represents the whole system as a cascade process. The learning algorithms, the theoretical analysis, and the results of real operation of a wastewater treatment plant are given to illustrate the efficiency of the approach.

Chapter 2 shows the design of a novel neural network that uses the geometric algebra framework and a bio-inspired optimization algorithm for the development of the training algorithm. The proposal, named hyperellipsoidal neural network (HNN), was applied to time series forecasting, showing high-accuracy results in comparison with classic approaches.

Chapter 3 introduces one of the most-used applications of neural networks: different approaches for modeling and control design of dynamical systems. The chapter illustrates how neural networks can be used to deal with complex and nonlinear systems, which are assumed to be uncertain, disturbed, unknown, or difficult to model. The chapter covers both the discrete-time and the continuous-time framework for modeling and control of uncertain systems, where recurrent neural networks (RNNs) are used to develop the artificial neural models, which are a posteriori employed for the design of two neural control schemes (sliding mode block control and nonlinear optimal control). Finally, the applicability of the neural methodology is illustrated through two practical examples.

Unmanned aerial vehicles (UAVs) have several capabilities that represent important advantages over another kinds of mobile robots, like ground vehicles. Flying robots have gained great interest in recent decades because of their ability to move in three-dimensional space and travel longer distances in less time. UAVs have important applications for both military and civil markets, such as rescue operations and surveillance. Therefore, Chapters 4 and 5 present two proposals to control UAVs using neural networks. In Chapter 4 a continuous-time decentralized neural control of a quadrotor UAV for trajectory tracking is proposed. A recurrent high-order neural network (RHONN) is used for online identification of the dynamics of the UAV; this is done using a series-parallel configuration and the filtered error learning law. Then, using the RHONN identification information, a local neural controller is derived via the backstepping approach. On the other hand, in Chapter 5 an adaptive controller based on the PID principle is shown using a neural network, which has shown adaptation capabilities. A multilayer perceptron is trained with the extended Kalman filter, and the output of the network represents the system control input. In both chapters, simulation results are provided in order to show the performance of the overall neural identification and control schemes.

In Chapter 6 an interesting application of neural network regression and interpolation is presented. An offline application to increase video quality using support vector regression (SVR) is proposed. The SVR is used as a pixel filter to reduce noise and increase the resolution and frame rate of video sequences.

In Chapter 7 the application and analysis of high-order artificial neural networks on bioprocess modeling and states prediction to overcome process constraints are shown. High-order neural network structures for an aerobic wastewater digestion process with organic compounds reduction and an anaerobic digestion process with biofuel production are proposed with the objective of modeling and predicting their dynamics. Two models are presented. In the first model, a recurrent high-order neural observer structure is proposed to estimate the complex states of the aerobic digestion process related to the effluent quality at the output. An analysis of the organic compounds reduction at the system output in the presence of disturbances is presented. In the second model, a recurrent neural network with autoregressive external input structure is proposed to predict the complex anaerobic digestion states related to biofuel production. In both cases, the neural network structure is trained with an extended Kalman filter algorithm. Simulation results illustrate the applicability of the recurrent neural network structures to estimate complex states of the biological processes in the presence of disturbances.

An important application related to disease control is presented in Chapter 8. RHONNs trained with an algorithm based on the extended Kalman filter are presented to identify dynamics of infectious diseases such as influenza A virus (IAV) and HIV. A negligible identification error is reported for both identifiers, showing that RHONNs are able to identify the within-host dynamics. Simulation results indicate promising directions for the problem of model identification of infectious diseases, serving for future model-based control strategies of infections.

Chapter 9 deals with the problem of attack detection and estimation for the type of cyber-physical systems known as networked control systems (NCSs). In NCSs, both the communication links and the physical system are vulnerable to a variety of attacks, and hence, it is of utmost importance to detect them and mitigate their effect on the system. By using a state-space representation of the communication network, controllers are designed to stabilize the traffic flow in the network and the NCS by using Q-learning in the case of linear NCSs and neural networks in the case of nonlinear

NCSs. The attack detection and estimation are accomplished by using residual signals that are generated from novel adaptive observers. Also, the detectability conditions are derived, and the maximum delay and packet loss bounds are given before the NCS becomes unstable. Simulation results are given.

A sensitivity analysis with artificial neural networks for the operation of photovoltaic systems is presented in Chapter 10. This analysis brings light to the interpretation of sensitivity of artificial neural networks' hidden layers. Sensitivity analysis represents a powerful tool that allows us to solve this problem, granting new skills to artificial neural network models that can be exploited in various areas, ranging from decision making in operating processes to optimization. A case of study of temperature estimation in photovoltaic modules is shown.

Chapter 11 presents a new approach to identify, classify, and control biomechatronic systems, which are controlled via superficial electromyographic signals generated by the upper limbs. Electromyographic data are recorded while the hand of subjects is constricted to grasps a set of spheres with a small variation in diameter. Five muscles are monitored with noninvasive electrodes placed on the skin of volunteers while a set of grasp–hold–relax tasks are carried out randomly. A pattern recognition module is used to classify data and to assign the extracted features to categories corresponding to each sphere grasped. A tracking generator is proposed using artificial neural networks, which are trained to learn the dynamics of finger motions. The performance of the methodology is evaluated in simulations and via real-time implementation with an embedded system.

Acknowledgment

The authors thank the National Council of Sciences and Technology (CONACYT), Mexico, for financially supporting the following projects: CB-256769, CB-256880, and PN-2016-4107. They also thank the University Center of Exact Sciences and Engineering of the University of Guadalajara (CUCEI-UDG), Mexico, for the support provided to write this book.

Nancy Arana-Daniel would like to give special thanks to her colleagues Carlos Villaseñor Padilla, Julio Esteban Valdes Lopez, Gehova Lopez Gonzalez, and Roberto Valencia Murillo, who have contributed their work, inspirations, and enriching discussions to the realization of this book.

Alma Y. Alanis is also grateful for the support given by "Fundación Marcos Moshinsky." In addition, her gratitude is extended to Edgar N. Sanchez and Jorge D. Rios who have contributed in different ways to the composition of this book.

Carlos Lopez-Franco thanks José de Jesús Hernández and Javier Gomez.

Alma Y. Alanis, Nancy Arana-Daniel, and
Carlos López-Franco

Contents

Hierarchical Dynamic Neural Networks for Cascade System Modeling With Application to Wastewater Treatment

WEN YU, DSC • DANIEL CARRILLO, DSC

1.1 INTRODUCTION

The input–output relation within a cascade process is very complex. It usually can be described by several non-linear subsystems, as is for example the case with the cascade process of wastewater treatment. Obviously, the first block and the last block cannot represent the whole process. Hierarchical models can be used to model this problem. When the cascade process is unknown, only the input and output data are available. Black-box modeling techniques are needed. Also, the internal variables of the cascade process need to be estimated.

There are three different approaches that can be used to model a cascade process. If the input/output data in each subblock are available, each model is identified independently. If the internal variables are not measurable, a general method is to regard the whole process as one block and to use one model to identify it [6,3,19]. Another method is to use hierarchical models to identify cascade processes. Advantages of this approach are that the cascade information is used for identification and that the internal variable can be estimated. In [2], discrete-time feedforward neural networks are applied to approximate the uncertainty parts of the cascade system.

Neural networks can approximate any nonlinear function to any prescribed accuracy provided a sufficient number of hidden neurons can be incorporated. Hierarchical neural models consisting of a number of low-dimensional neural systems have been presented by [9] and [11] in order to avoid the dimension explosion problem. The main applications of hierarchical models are fuzzy systems, because rule explosion problem can be avoided in hierarchical systems [9], for example, in hierarchical fuzzy neural network [17], hierarchical fuzzy systems [12], and hierarchical fuzzy cerebellar model articulation controller (CMAC) networks [15]. Sensitivity analysis of the hierarchical fuzzy model was given in [12]. A statistical learning method was employed to construct hierarchical models in [3]. Based on Kolmogorov's theorem, [18] showed that any continuous function can be represented as a superposition of functions with the natural hierarchical structure. In [16], fuzzy CMAC networks are formed into a hierarchical structure.

The normal training method of hierarchical neural systems is still gradient descent. The key for the training of hierarchical neural models is to get an explicit expression of each internal error. Normal identification algorithms (gradient descent, least square, etc.) are stable under ideal conditions. They might become unstable in the presence of unmodeled dynamics. The Lyapunov approach can be used directly to obtain robust training algorithms of continuous-time and discrete-time neural networks. By using passivity theory, [5], [8], and [14] successfully proved that gradient descent algorithms of continuous-time dynamic neural networks were stable and robust to any bounded uncertainties.

The main problem for the training of a hierarchical neural model is the estimation of the internal variable. In this chapter, a hierarchical dynamic neural network is applied to model wastewater treatment. Two stable training algorithms are proposed. A novel approximate method for the internal variable of the cascade process is discussed. Real application results show that the new modeling approach is effective for this cascade process.

1.2 CASCADE PROCESS MODELING VIA HIERARCHICAL DYNAMIC NEURAL NETWORKS

Each subprocess of a cascade process, such as wastewater treatment, can be described using the following general nonlinear dynamic equation:

$$\dot{x} = f(x, u), \qquad (1.1)$$

where $x \in R^{n_c}$ is the inner state, $u \in R^{m_c}$ is the input, and f is a vector function. Without loss of generality,

Artificial Neural Networks for Engineering Applications. https://doi.org/10.1016/B978-0-12-818247-5.00010-1

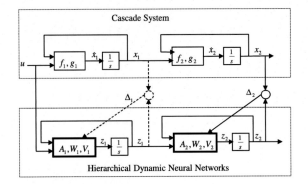

FIG. 1.1 Cascade system modeling via hierarchical recurrent neural networks.

we use two nonlinear affine systems to show how to use the hierarchical dynamic neural networks to model the system; see Fig. 1.1. The identified cascade nonlinear systems are given by

$$\begin{aligned} \dot{x}_1 &= f_1(x_1) + g_1(x_1)\,u, \\ \dot{x}_2 &= f_2(x_2) + g_2(x_2)\,x_1, \end{aligned} \tag{1.2}$$

where $x_1, x_2 \in R^n$ are the inner states of the subsystems, and f_1, f_2, g_1, and g_2 are unknown vector functions; x_1 can be also regarded as the output of subsystem 1 and as the input of subsystem 2; $u \in R$ is the input of the whole system, and also the input of subsystem 1; x_2 is the output of the whole system, and also the output of subsystem 2.

Only u and x_2 are available for the cascade process modeling. Since the internal variables are not measurable, a general method is to regard the whole process as one block and to use one model to identify it. Another method is to use hierarchical models to identify cascade processes. The advantages of this approach are that the cascade information is used for identification and that the internal variable can be estimated. In Section 1.3 we will show how to approximate it. In many wastewater treatment plants, x_1 is sampled occasionally. We can use this real value to improve the modeling accuracy.

We construct the following hierarchical dynamic neural networks to model (1.2):

$$\begin{aligned} \dot{z}_1 &= A_1 z_1 + W_1 \sigma_1(z_1) + V_1 \phi_1(z_1)u, \\ \dot{z}_2 &= A_2 z_2 + W_2 \sigma_2(z_2) + V_2 \phi_2(z_2)z_1, \end{aligned} \tag{1.3}$$

where $z_1, z_2 \in R^n$ are the states of the neural models, and W_1, W_2, V_1, and V_2 are the weights of the neural networks; A_1 and A_2 are known stable matrices. The active functions of $\sigma(\cdot)$ and $\phi(\cdot)$ are used as sigmoid

functions, i.e.,

$$\sigma(z) = \frac{a}{1 + e^{-bz}} - c.$$

From Fig. 1.1, the modeling error is

$$\Delta_2 = z_2 - x_2. \tag{1.4}$$

The internal modeling error is

$$\Delta_1 = z_1 - x_1. \tag{1.5}$$

This will be estimated.

Generally, the hierarchical dynamic neural networks (1.3) cannot follow the nonlinear system (1.2) exactly. The nonlinear cascade system may be written as

$$\begin{aligned} \dot{x}_1 &= A_1 x_1 + W_1^* \sigma_1(x_1) + V_1^* \phi_1(x_1)u - \tilde{f}_1, \\ \dot{x}_2 &= A_2 x_2 + W_2^* \sigma_2(x_2) + V_2^* \phi_2(x_2)z_1 - \tilde{f}_2, \end{aligned} \tag{1.6}$$

where W_1^*, W_2^*, V_1^*, and V_2^* are unknown bounded matrices. We assume the upper bounds, $\bar{W}_1$, $\bar{W}_2$, $\bar{V}_1$, and $\bar{V}_2$, are known as

$$\begin{aligned} &W_1^* \Lambda_w^{-1} W_1^{*T} \le \bar{W}_1, \quad W_2^* \Lambda_w^{-1} W_2^{*T} \le \bar{W}_2, \\ &\Lambda_w = \Lambda_w^T > 0, \\ &V_1^* \Lambda_v^{-1} V_1^{*T} \le \bar{V}_1, \quad W_2^* \Lambda_v^{-1} W_2^{*T} \le \bar{V}_2, \\ &\Lambda_v = \Lambda_v^T > 0, \end{aligned} \tag{1.7}$$

where $\tilde{f}_1$ and $\tilde{f}_2$ are modeling errors and disturbances. Since the state and output variables are physically bounded, the modeling errors $\tilde{f}_1$ and $\tilde{f}_2$ can be assumed to be bounded too. The upper bounds of the modeling errors are

$$\left\| \tilde{f}_1 \right\|_{\Lambda_{f_1}}^2 \le \bar{\eta}_{f_1} \le \infty, \quad \left\| \tilde{f}_2 \right\|_{\Lambda_{f_2}}^2 \le \bar{\eta}_{f_2} \le \infty, \tag{1.8}$$

where $\bar{\eta}_{f_1}$ and $\bar{\eta}_{f_2}$ are known positive matrices, and Λ_{f_1} and Λ_{f_2} are any positive definite matrices.

Now we calculate the following errors:

$$\begin{aligned} &W_1 \sigma_1(z_1) - W_1^* \sigma_1(x_1) \\ &= \left[W_1 \sigma_1(z_1) - W_1^* \sigma_1(z_1) \right] + \left[W_1^* \sigma_1(z_1) - W_1^* \sigma_1(x_1) \right] \\ &= \tilde{W}_1 \sigma_1(z_1) + W_1^* \left[\sigma_1(z_1) - \sigma_1(x_1) \right] \\ &= \tilde{W}_1 \sigma_1(z_1) + W_1^* \tilde{\sigma}_1, \end{aligned} \tag{1.9}$$

where $\tilde{W}_1 = W_1 - W_1^*$, $\tilde{\sigma}_1 = \sigma_1(z_1) - \sigma_1(x_1)$. Similarly,

$$\begin{aligned} &W_2 \sigma_2(z_2) - W_2^* \sigma_2(x_2) = \tilde{W}_2 \sigma_2(z_2) + W_2^* \tilde{\sigma}_2, \\ &V_1 \sigma_1(z_1)u - V_1^* \phi_1(x_1)u = \tilde{V}_1 \sigma_1(z_1)u + V_1^* \tilde{\phi}_1 u, \\ &V_2 \phi_2(z_2)z_1 - V_2^* \phi_2(x_2)z_1 = \tilde{V}_2 \sigma_2(z_2)z_1 + V_2^* \tilde{\phi}_2 z_1. \end{aligned}$$

Because $\sigma(\cdot)$ and $\phi(\cdot)$ are chosen as sigmoid functions, they satisfy the following Lipschitz property:

$$\tilde{\sigma}_1^T \Lambda_w \tilde{\sigma}_1 \leq \Delta_1^T D_{\sigma 1} \Delta_1, \quad \tilde{\phi}_1^T \Lambda_v \tilde{\phi}_1 \leq \Delta_1^T D_{\phi 1} \Delta_1,$$
$$\tilde{\sigma}_2^T \Lambda_w \tilde{\sigma}_2 \leq \Delta_2^T D_{\sigma 2} \Delta_2, \quad \tilde{\phi}_2^T \Lambda_v \tilde{\phi}_2 \leq \Delta_2^T D_{\phi 2} \Delta_2,$$
$$(1.10)$$

where Λ_w, Λ_v, $D_{\sigma 1}$, $D_{\phi 1}$, $D_{\sigma 2}$, and $D_{\phi 2}$ are positive definite matrices.

1.3 STABLE TRAINING OF THE HIERARCHICAL DYNAMIC NEURAL NETWORKS

In order to obtain a stable training algorithm for the hierarchical dynamic neural networks (1.3), we calculate the error dynamics of the submodels. From (1.3) and (1.6), we have

$$\dot{\Delta}_1 = A_1 \Delta_1 + \tilde{W}_1 \sigma_1(z_1) + \tilde{V}_1 \phi_1(z_1)u + W_1^* \tilde{\sigma}_1$$
$$+ V_1^* \tilde{\phi}_1 u + \tilde{f}_1,$$
$$\dot{\Delta}_2 = A_2 \Delta_2 + \tilde{W}_2 \sigma_2(z_2) + \tilde{V}_2 \phi_2(z_2)z_1 + W_2^* \tilde{\sigma}_2$$
$$+ V_2^* \tilde{\phi}_2 z_1 + \tilde{f}_2.$$
$$(1.11)$$

If the outputs of all blocks are available, we can train each block independently via the modeling errors between neural models and the corresponding process blocks, Δ_1 and Δ_2. Let us define

$$R_1 = \bar{W}_1 + \bar{V}_1, \quad Q_1 = (D_{\sigma 1} + \|u\| D_{\phi 1} + Q_{10}) \quad (1.12)$$

and the matrices A_1 and Q_{10} are selected to fulfill the following conditions:

(1) the pair $(A_1, R_1^{1/2})$ is controllable, the pair $(Q_1^{1/2}, A_1)$ is observable;

(2) if the local frequency condition [1] is satisfied, i.e.,

$$A_1^T R_1^{-1} A_1 - Q_1$$
$$\geq \frac{1}{4} \left[A_1^T R_1^{-1} - R_1^{-1} A_1 \right] R_1 \left[A_1^T R_1^{-1} - R_1^{-1} A_1 \right]^T,$$
$$(1.13)$$

then the following assumption can be established.

A1: There exist a stable matrix A_1 and a strictly positive definite matrix Q_{10} such that the matrix Riccati equation

$$A_1^T P_1 + P_1 A_1 + P_1 R_1 P_1 + Q_1 = 0 \quad (1.14)$$

has a positive solution $P_1 = P_1^T > 0$.

This condition is easily fulfilled if we select A_1 as a stable diagonal matrix. The next theorem states the learning procedure of a neuroidentifier. Similarly, there exist a stable matrix A_2 and a strictly positive definite matrix Q_{20} such that the matrix Riccati equation

$$A_2^T P_2 + P_2 A_2 + P_2 R_2 P_2 + Q_2 = 0, \quad (1.15)$$

where $R_2 = \bar{W}_2 + \bar{V}_2$, $Q_2 = (D_{\sigma 2} + \|z_1\| D_{\phi 2} + Q_{20})$.

First, we may choose A_1 and Q_1 such that the Riccati equation (1.14) has a positive solution P_1. Then Λ_{f1} may be found according to the condition (1.8). Since (1.7) is correct for any positive definite matrix, (1.8) can be established if Λ_{f1} is selected as a small enough constant matrix. The condition (1.8) has no effect on the network dynamics (1.3) and its training (1.16).

Theorem 1.1. *If Assumption A1 is satisfied and the weights* $W_{1,t}$ *and* $W_{2,t}$ *are updated as*

$$\frac{d}{dt}\left(\tilde{W}_2^T\right) = -K_{w2} P_2 \Delta_2 \sigma_2^T (z_2),$$
$$\frac{d}{dt}\left(\tilde{V}_2^T\right) = -K_{v2} P_2 \Delta_2 \left[\phi_2(z_2)z_1\right]^T,$$
$$\frac{d}{dt}\left(\tilde{W}_1^T\right) = -K_{w1} P_1 \Delta_1 \sigma_1^T (z_1),$$
$$\frac{d}{dt}\left(\tilde{V}_1^T\right) = -K_{v1} P_1 \Delta_1 \left[\phi_1(z_1)u\right]^T,$$
$$(1.16)$$

where P_1 *and* P_1 *are the solution of the Riccati equation (1.14) and (1.15), then the identification error dynamics (1.11) is strictly passive from the modeling error* $\tilde{f}_1$ *and* $\tilde{f}_2$ *to the identification errors* $2P_1\Delta_1$ *and* $2P_2\Delta_2$ *and the updating law (1.16) can make the identification procedure stable.*

Proof. Select a Lyapunov function (storage function) as

$$S = \Delta_1^T P_1 \Delta_1 + tr\left\{\tilde{W}_1^T K_{w1}^{-1} \tilde{W}_1\right\} + tr\left\{\tilde{V}_1^T K_{v1}^{-1} \tilde{V}_1\right\}$$
$$+ \Delta_2^T P_2 \Delta_2 + tr\left\{\tilde{W}_2^T K_{w2}^{-1} \tilde{W}_2\right\} + tr\left\{\tilde{V}_2^T K_{v2}^{-1} \tilde{V}_2\right\},$$
$$(1.17)$$

where $P \in \Re^{n \times n}$ is a positive definite matrix. According to (1.11), the derivative is

$$\dot{S} = \Delta_1^T \left(P_1 A_1 + A_1^T P_1\right)\Delta_1 + 2\Delta_1^T P_1 \tilde{W}_1 \sigma_1(z_1)$$
$$+ 2\Delta_1^T P_1 \tilde{V}_1 \phi_1(z_1)u + \Delta_2^T \left(P_2 A_2 + A_2^T P_2\right)\Delta_2$$
$$+ 2\Delta_2^T P_2 \tilde{W}_2 \sigma_2(z_2) + 2\Delta_2^T P_2 \tilde{V}_2 \phi_2(z_2)z_1$$
$$+ 2\Delta_1^T P_1 \left(W_1^* \tilde{\sigma}_1 + V_1^* \tilde{\phi}_1 u + \tilde{f}_1\right)$$
$$+ 2\Delta_2^T P_2 \left(W_2^* \tilde{\sigma}_2 + V_2^* \tilde{\phi}_2 z_1 + \tilde{f}_2\right)$$
$$+ 2tr\left\{\frac{d}{dt}\left(\tilde{W}_1^T\right)K_{w1}^{-1}\tilde{W}_1\right\} + 2tr\left\{\frac{d}{dt}\left(\tilde{W}_2^T\right)K_{w2}^{-1}\tilde{W}_2\right\}$$
$$+ 2tr\left\{\frac{d}{dt}\left(\tilde{V}_1^T\right)K_{v1}^{-1}\tilde{V}_1\right\} + 2tr\left\{\frac{d}{dt}\left(\tilde{V}_2^T\right)K_{v2}^{-1}\tilde{V}_2\right\}.$$

Since $2\Delta_1^T P_1 W_1^* \tilde{\sigma}_1$ is a scalar, using (1.10) and the matrix inequality

$$X^T Y + \left(X^T Y\right)^T \le X^T \Lambda^{-1} X + Y^T \Lambda Y, \qquad (1.18)$$

where $X, Y, \Lambda \in \Re^{n \times k}$ are any matrices and Λ is any positive definite matrix, we obtain

$$
\begin{aligned}
2\Delta_1^T P_1 W_1^* \tilde{\sigma}_1 &\le \Delta_1^T P_1 W_1^* \Lambda_w^{-1} W_1^{*T} P_1 \Delta_1 + \tilde{\sigma}_1^T \Lambda_w \tilde{\sigma}_1 \\
&\le \Delta_1^T \left(P_1 \bar{W}_1 P_1 + D_{\sigma 1}\right) \Delta_1.
\end{aligned}
\qquad (1.19)
$$

Similarly,

$$
\begin{aligned}
2\Delta_2^T P_2 W_2^* \tilde{\sigma}_2 &\le \Delta_2^T \left(P_2 \bar{W}_2 P_2 + D_{\sigma 2}\right) \Delta_2, \\
2\Delta_1^T P_1 V_1^* \tilde{\phi}_1 u &\le \Delta_1^T \left(P_1 \bar{V}_1 P_1 + \|u\| D_{\phi 1}\right) \Delta_1, \\
2\Delta_2^T P_2 V_2^* \tilde{\phi}_2 z_1 &\le \Delta_2^T \left(P_2 \bar{V}_2 P_2 + \|z_1\| D_{\phi 2}\right) \Delta_2.
\end{aligned}
$$

So we have

$$
\begin{aligned}
\dot{S} \le\ &\Delta_1^T \left[\begin{array}{c} P_1 A_1 + A_1^T P_1 + P_1 \left(\bar{W}_1 + \bar{V}_1\right) P_1 \\ + \left(D_{\sigma 1} + \|u\| D_{\phi 1} + Q_{10}\right) \end{array} \right] \Delta_1 \\
&- \Delta_1^T Q_{10} \Delta_1 \\
&+ \Delta_2^T \left[\begin{array}{c} P_2 A_2 + A_2^T P_2 + P_2 \left(\bar{W}_2 + \bar{V}_2\right) P_2 \\ + \left(D_{\sigma 3} + \|z_1\| D_{\phi 2} + Q_{20}\right) \end{array} \right] \Delta_2 \\
&- \Delta_2^T Q_{20} \Delta_2 \\
&+ 2tr \left\{ \frac{d}{dt} \left(\tilde{W}_2^T\right) K_{w2}^{-1} \tilde{W}_2 \right\} + 2\Delta_2^T P_2 \tilde{W}_2 \sigma_2(z_2) \\
&+ 2\Delta_2^T P_2 \tilde{f}_2 \\
&+ 2tr \left\{ \frac{d}{dt} \left(\tilde{V}_2^T\right) K_{v2}^{-1} \tilde{V}_2 \right\} + 2\Delta_2^T P_2 \tilde{V}_2 \phi_2(z_2) z_1 \\
&+ 2tr \left\{ \frac{d}{dt} \left(\tilde{W}_1^T\right) K_{w1}^{-1} \tilde{W}_1 \right\} + 2\Delta_1^T P_1 \tilde{W}_1 \sigma_1(z_1) \\
&+ 2\Delta_1^T P_1 \tilde{f}_1 \\
&+ 2tr \left\{ \frac{d}{dt} \left(\tilde{V}_1^T\right) K_{v1}^{-1} \tilde{V}_1 \right\} + 2\Delta_1^T P_1 \tilde{V}_1 \phi_1(z_1) u.
\end{aligned}
$$

Using (1.14), (1.15), (1.16), and $\frac{d}{dt}\left(\tilde{W}_1^T\right) = \frac{d}{dt} \tilde{W}_1$,

$$\dot{S} \le -\Delta_1^T Q_{10} \Delta_1 + 2\Delta_1^T P_1 \tilde{f}_1 - \Delta_2^T Q_{20} \Delta_2 + 2\Delta_2^T P_2 \tilde{f}_2. \qquad (1.20)$$

From the passivity definition, if we define the inputs as $\tilde{f}_1$ and $\tilde{f}_2$ and the outputs as $2P_1 \Delta_1$ and $2P_2 \Delta_2$, then the system is strictly passive with $\Delta_1^T Q_{10} \Delta_1 \ge 0$ and $\Delta_2^T Q_{20} \Delta_2$.

In view of the matrix inequality (1.18),

$$
\begin{aligned}
2\Delta_1^T P_1 \tilde{f}_1 &\le \Delta_1^T P_1 \Lambda_{f1} P_1 \Delta_1 + \tilde{f}_1^T \Lambda_{f1}^{-1} \tilde{f}_1, \\
2\Delta_2^T P_2 \tilde{f}_2 &\le \Delta_2^T P_2 \Lambda_{f2} P_2 \Delta_2 + \tilde{f}_2^T \Lambda_{f2}^{-1} \tilde{f}_2,
\end{aligned}
\qquad (1.21)
$$

(1.20) can be represented as

$$
\begin{aligned}
\dot{S}_t \le\ &-\lambda_{\min}(Q_{10}) \|\Delta_1\|^2 - \lambda_{\min}(Q_{20}) \|\Delta_2\|^2 \\
&+ \Delta_1^T P_1 \Lambda_{f1} P_1 \Delta_1 + \tilde{f}_1^T \Lambda_{f1}^{-1} \tilde{f}_1 + \Delta_2^T P_2 \Lambda_{f2} P_2 \Delta_2 \\
&+ \tilde{f}_2^T \Lambda_{f2}^{-1} \tilde{f}_2 \\
\le\ &\left(-\alpha_{\|\Delta_1\|} \|\Delta_1\| + \beta_{\left\|\tilde{f}_1\right\|} \left\|\tilde{f}_1\right\| \right) \\
&+ \left(-\alpha_{\|\Delta_2\|} \|\Delta_2\| + \beta_{\left\|\tilde{f}_2\right\|} \left\|\tilde{f}_2\right\| \right),
\end{aligned}
$$

where $\alpha_{\|\Delta_1\|} = \left[\lambda_{\min}(Q_{10}) - \lambda_{\max}\left(P_1 \Lambda_{f1} P_1\right)\right] \|\Delta_1\|$, $\beta_{\left\|\tilde{f}_1\right\|} = \lambda_{\max}\left(\Lambda_{f1}^{-1}\right) \left\|\tilde{f}_1\right\|$, $\alpha_{\|\Delta_2\|} = \left[\lambda_{\min}(Q_{20}) - \lambda_{\max}\left(P_2 \Lambda_{f2} P_2\right)\right] \|\Delta_2\|$, and $\beta_{\left\|\tilde{f}_2\right\|} = \lambda_{\max}\left(\Lambda_{f2}^{-1}\right) \left\|\tilde{f}_2\right\|$. We can select positive definite matrices Λ_{f1} and Λ_{f2} such that (1.8) is established. So $\alpha_{\|\Delta_1\|}$, $\alpha_{\|\Delta_2\|}$, $\beta_{\left\|\tilde{f}_1\right\|}$, and $\beta_{\left\|\tilde{f}_2\right\|}$ are $\mathcal{K}_\infty$ functions and S_t is an input-to-state stable (ISS)-Lyapunov function. The dynamics of identification error (1.11) is ISS. So when the modeling errors $\tilde{f}_1$ and $\tilde{f}_2$ are bounded, the updating law (1.16) can make the modeling errors stable, i.e.,

$$\Delta_1 \in L_\infty, \qquad \Delta_2 \in L_\infty. \qquad \square$$

Since the updating rates in (1.16) are $K_i P_j$, and K_i can be selected as any positive matrix, the learning process of the dynamic neural network is free of the solution of the Riccati equation (1.14).

Theorem 1.2. *If the modeling errors $\tilde{f}_1 = \tilde{f}_2 = 0$ (only uncertainty parameters present), then the updating law (1.16) can make the identification error asymptotically stable, i.e.,*

$$\lim_{t \to \infty} \Delta_1 = 0, \qquad \lim_{t \to \infty} \Delta_2 = 0. \qquad (1.22)$$

Proof. The modeling errors $\tilde{f}_1 = \tilde{f}_2 = 0$, $2\Delta_1^T P_1 \tilde{f}_1 = 2\Delta_2^T P_2 \tilde{f}_2 = 0$, and the storage function (1.20) satisfies

$$\dot{S} \le -\Delta_1^T Q_{10} \Delta_1 - \Delta_2^T Q_{20} \Delta_2 \le 0.$$

The positive definite $S(x_t)$ implies Δ_1, Δ_2 and the weights are bounded. From the error equation (1.11) $\dot{\Delta}_1 \in L_\infty$, $\dot{\Delta}_2 \in L_\infty$. Integrate (1.20) on both sides to obtain

$$\int_0^\infty \|\Delta_1\|_{Q_{10}} + \|\Delta_2\|_{Q_{20}} \le S_0 - S_\infty < \infty.$$

So $\Delta_1 \in L_2 \cap L_\infty$, $\Delta_2 \in L_2 \cap L_\infty$, and using Barbalat's lemma, we have (1.22). Since u, σ, ϕ, and P are bounded, $\lim\limits_{t\to\infty} \Delta_1 = 0$, $\lim\limits_{t\to\infty} \Delta_2 = 0$. □

For many cascade systems the outputs in the internal blocks are not measurable, for example Δ_1. The modeling error in the final block Δ_2 should be propagated to the other blocks, i.e., we should calculate the internal modeling error Δ_1 from Δ_2.

From (1.3) and (1.6), the last block can be written as

$$\begin{aligned} \dot{z}_2 &= A_2 z_2 + W_2 \sigma_2(z_2) + V_2 \phi_2(z_2) z_1, \\ \dot{x}_2 &= A_2 x_2 + W_2 \sigma_2(x_2) + V_2 \phi_2(z_2) x_1 - \tilde{f}_3, \end{aligned} \quad (1.23)$$

where $\tilde{f}_3$ is the modeling error corresponding to the weights W and V_2. So

$$\dot{\Delta}_2 = A_2 \Delta_2 + W_2 \tilde{\sigma}_2 + V_2 \phi_2(z_2) \Delta_1 + \tilde{f}_3.$$

So Δ_1 is approximated by

$$\Delta_1 = \left[V_2 \phi_2(z_2) \right]^{-1} \left[\dot{\Delta}_2 - A_2 \Delta_2 - W_2 \tilde{\sigma}_2 - \tilde{f}_3 \right].$$

In Fig. 1.1 we note $\hat{f}_3$ is the difference between the block "f_2, g_2" and the block "A_2, W_2, V_2." So $\hat{f}_3$ can be estimated as

$$\tilde{f}_3 \approx \dot{z}_2 - \dot{\Delta}_2.$$

Here

$$\dot{z}_2 = \frac{z_2(t) - z_2(t-\tau)}{\tau} + \delta(t), \quad \tau > 0,$$

where $\delta(t)$ is the differential approximation error. So

$$\tilde{f}_3 \approx \hat{f}_3 = \frac{z_2(t) - z_2(t-\tau)}{\tau} - \frac{\Delta_2(t) - \Delta_2(t-\tau)}{\tau}.$$

The internal modeling error is approximated by

$$\begin{aligned} \hat{\Delta}_1 = &\left[V_2 \phi_2(z_2) \right]^{-1} \\ &\times \left[\begin{array}{c} 2\frac{\Delta_2(t)-\Delta_2(t-\tau)}{\tau} - \frac{z_2(t)-z_2(t-\tau)}{\tau} \\ -A_2 \Delta_2 - W_2 \left[\sigma_2(z_2) - \sigma_2(x_2) \right] \end{array} \right]. \end{aligned} \quad (1.24)$$

In order to ensure $\hat{\Delta}_1$ is bounded, we use (1.24) when $\|V_2 \phi_2(z_2)\| > \tau$. Otherwise, we use Δ_2 to represent Δ_1, i.e., $\hat{\Delta}_1 = \Delta_2$.

Although the gradient algorithm (1.16) can ensure the modeling errors Δ_1 and Δ_2 are bounded (Theorem 1.1), the structure uncertainties $\tilde{f}_1$ and $\tilde{f}_2$ will cause the parameters drift for the gradient algorithm (1.16). Some robust modification should be applied to make

the parameters (weights) stable. In order to guarantee the overall models are stable, we use the following dead-zone training algorithm.

Theorem 1.3. *The weights are adjusted as follows:*

(a) if $\|\Delta_1\|^2 > \frac{\bar{\eta}_{f_1}}{\lambda_{\min}(Q_1)}$ and $\|\Delta_2\|^2 > \frac{\bar{\eta}_{f_2}}{\lambda_{\min}(Q_2)}$, then the updating law is given by (1.16);

(b) if $\|\Delta_1\|^2 \le \frac{\bar{\eta}_{f_1}}{\lambda_{\min}(Q_1)}$ or $\|\Delta_2\|^2 > \frac{\bar{\eta}_{f_2}}{\lambda_{\min}(Q_2)}$, then we stop the learning procedure (all right-hand sides of the corresponding system of differential equations are equal to zero) and maintain all weights constant; then, besides the modeling errors being bounded, the weight matrices also remain bounded and for any $T > 0$ the identification error fulfills the following tracking performance:

$$\lim_{T\to\infty} \frac{1}{T} \int_0^T \left(\|\Delta_1\|_{Q_1}^2 + \|\Delta_2\|_{Q_2}^2 \right) dt \le \kappa_1 \bar{\eta}_{f_1} + \kappa_2 \bar{\eta}_{f_2}, \quad (1.25)$$

where κ is the condition number of Q defined as $\kappa_1 = \frac{\lambda_{\max}(Q_1)}{\lambda_{\min}(Q_1)}$, $\kappa_2 = \frac{\lambda_{\max}(Q_2)}{\lambda_{\min}(Q_2)}$.

Proof. From (1.21), (1.14), and (1.15), (1.20) can be rewritten as

$$\begin{aligned} \dot{S}_t &\le -\Delta_1^T Q_1 \Delta_1 - \Delta_2^T Q_2 \Delta_2 + \tilde{f}_1^T \Lambda_{f1}^{-1} \tilde{f}_1 + \tilde{f}_2^T \Lambda_{f2}^{-1} \tilde{f}_2 \\ &\le -\Delta_1^T Q_1 \Delta_1 - \Delta_2^T Q_2 \Delta_2 + \bar{\eta}_{f_1} + \bar{\eta}_{f_2}. \end{aligned} \quad (1.26)$$

(I) If $\|\Delta_1\|^2 > \frac{\bar{\eta}_{f_1}}{\lambda_{\min}(Q_1)}$ and $\|\Delta_2\|^2 > \frac{\bar{\eta}_{f_2}}{\lambda_{\min}(Q_2)}$, using the updating law (1.16) we conclude that $\dot{S}_t < 0$. S_t is bounded. Integrating (1.26) from 0 to T yields

$$\begin{aligned} S_T - S_0 \\ \le -\int_0^T \left(\Delta_1^T Q_1 \Delta_1 + \Delta_2^T Q_2 \Delta_2 \right) dt + \left(\bar{\eta}_{f_1} + \bar{\eta}_{f_2} \right) T. \end{aligned}$$

Because $\kappa \ge 1$ and $S_T \ge 0$, we have

$$\begin{aligned} \int_0^T \left(\Delta_1^T Q_1 \Delta_1 + \Delta_2^T Q_2 \Delta_2 \right) dt &\le S_0 + \left(\bar{\eta}_{f_1} + \bar{\eta}_{f_2} \right) T \\ &\le S_0 + \left(\kappa_1 \bar{\eta}_{f_1} + \kappa_2 \bar{\eta}_{f_2} \right) T, \end{aligned} \quad (1.27)$$

where κ_1 is the condition number of Q_1.

(II) If $\|\Delta_1\|^2 > \frac{\bar{\eta}_{f_1}}{\lambda_{\min}(Q_1)}$ or $\|\Delta_2\|^2 > \frac{\bar{\eta}_{f_2}}{\lambda_{\min}(Q_2)}$, the weights become constants and S_t remains bounded.

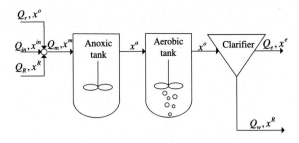

FIG. 1.2 Wastewater treatment process.

Since $S_0 \geq 0$,

$$\int_0^T \left(\Delta_1^T Q_1 \Delta_1 + \Delta_2^T Q_2 \Delta_2 \right) dt$$
$$\leq \int_0^T \left(\lambda_{\max}(Q_1) \|\Delta_1\|^2 + \lambda_{\max}(Q_2) \|\Delta_2\|^2 \right) dt$$
$$\leq \left(\frac{\lambda_{\max}(Q_1)}{\lambda_{\min}(Q_1)} \bar{\eta}_{f_1} + \frac{\lambda_{\max}(Q_2)}{\lambda_{\min}(Q_2)} \bar{\eta}_{f_2} \right) T$$
$$\leq S_0 + \left(\kappa_1 \bar{\eta}_{f_1} + \kappa_2 \bar{\eta}_{f_2} \right) T. \tag{1.28}$$

From (I) and (II), S_t is bounded. Because $W_{1,0} = W_1^*$, $W_{2,0} = W_2^*$, $V_{1,0} = V_1^*$ and $V_{2,0} = V_2^*$. From (1.27) and (1.28), (1.25) is obtained. The theorem is proved. $\square$

1.4 MODELING OF WASTEWATER TREATMENT

The wastewater treatment plant studied in this chapter is an anoxic/oxidant nitrogenous removal process [10]. It consists of two biodegradation tanks and a secondary clarifier in series form; see Fig. 1.2. Here, Q_{in}, Q_m, Q_r, Q_R, and Q_w denote flow of wastewater to be disposed, flow of mixed influent, flow of internal recycles, flow of external recycles, and flow of surplus sludge, respectively. Water quality indices such as chemical oxygen demand (COD), biological oxygen demand (BOD$_5$), NH4-N (ammonia), nitrate, and suspended solid (SS) are decomposed into those components in activated sludge models (ASMs) [4]. The state variable $\mathbf{x}$ is defined as

$$\mathbf{x} = [S_I, S_S, X_I, X_S, X_P, X_{BH}, X_{BA},$$
$$S_{NO}, S_{NH}, S_O, S_{ND}, X_{ND}, S_{alk}]^T, \tag{1.29}$$

where S_I is soluble inert and S_S is readily biodegradable substrate, X_I is suspended inert and X_S is slowly biodegradable substrate, X_P is suspended inert products, X_{BH} is autotrophic biomass, X_{BA} is heterotrophic biomass, S_{NO} is nitrate, S_{NH} is ammonia, S_O is soluble oxygen, S_{ND} is soluble organic nitrogen, X_{ND} is suspended organic nitrogen, and S_{alk} is alkalinity. In this

tank, there are three major reaction processes, i.e.,

$$NH_4^+ + 1.5O_2 \rightarrow NO_2^- + H_2O + 2H^+,$$
$$NO_2^- + 0.5O_2 \rightarrow NO_3^-, \tag{1.30}$$
$$COD + O_2 \rightarrow CO_2 + H_2O + AS,$$

where COD represents carbonous contamination and AS denotes activated sludge. Nitrite is recycled from the aerobic tanks. It is deoxidized by autotrophic microorganisms in the denitrification phase by the following reaction:

$$2NO_3^- + 2H^+ \rightarrow N_2 + H_2O + 2.5O_2. \tag{1.31}$$

The water quality index COD depends on the control input

$$\mathbf{w} = [w_i] = [Q_{in}, Q_r, Q_R, Q_w]^T \tag{1.32}$$

and on the influent quality $\mathbf{x}^{in}$, i.e.,

$$COD = f(Q_{in}, Q_r, Q_R, Q_w, \mathbf{x}^{in}).$$

COD is also affected by other external factors such as temperature, flow distribution, and toxins. It is very difficult to find the nonlinear function $f(\cdot)$ [13].

We know each biological reactor in wastewater treatment plants can be described by the following dynamic equation:

$$\dot{\mathbf{x}}(t) = A\mathbf{x}(t) + B\mathbf{x}_b(t) + \varphi(\mathbf{x}(t)), \tag{1.33}$$

where $\mathbf{x} \in R^{13}$ is the inner state, which is defined in (1.29), $\mathbf{x}_b(t) \in R^4$ is the input, which is defined in (1.32), φ denotes the reaction rates, $\varphi(\cdot) \in R^{13}$, $A = -\frac{w_1(t)+w_2(t)+w_3(t)}{V}$, and $B = \frac{w_1(t)+w_2(t)+w_3(t)}{V}$.

The resulting steady-state values of anoxic and aerobic reactors are shown in Table 1.1. Used data are from 2003.

As discussed above, hierarchical dynamic neural networks are suitable for modeling this process. Each neural network corresponds to a reaction rate in a reactor, i.e.,

$$\dot{z}_1 = A_1 z_1 + G_1 \sigma_1(z_1) + H_1 \mathbf{x}_u,$$
$$\dot{z}_2 = A_2 z_2 + G_2 \sigma_2(z_2) + H_2 \phi_2(z_2) z_1.$$

The design choices are as follows: both neural networks have one hidden layer, each hidden layer has 50 hidden nodes. The training algorithms of each neural model are (1.16); here the activation function $\phi_i(\cdot) = \tanh(x) = \frac{e^x - e^{-x}}{e^x + e^{-x}}$, $\eta_0 = 1$, and the initial weights of $W(1)$ and $V(1)$ are random numbers between [0, 1].

Dynamic modeling uses the steady-state values resulting from steady-state simulations as initial values with a hydraulic residence time of 10.8 h and a sludge

TABLE 1.1
Steady-state values of anoxic and aerobic reactors.

	S_S	X_{BH}	X_S	X_I	S_{NH}	S_I	S_{ND}	X_{ND}	S_O	X_{BA}	S_{NO}	X_P	S_{alk}
Anoxic	1.2518	3249	74.332	642.4	7.9157	38.374	0.7868	5.7073	0.0001	220.86	3.9377	822.19	4.9261
Aerobic	0.6867	3244.8	47.392	643.36	0.1896	38.374	0.6109	3.7642	1.4988	222.39	12.819	825.79	3.7399

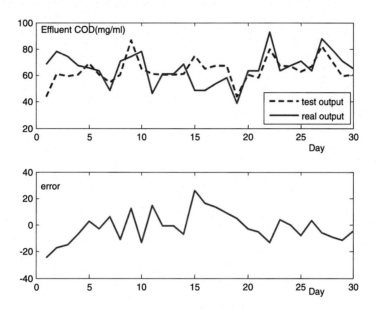

FIG. 1.3 The results using hierarchical dynamic neural networks.

TABLE 1.2
Comparison results of an activated sludge model (ASM), linear model (LM), neural network (NN), and hierarchical neural network (HNN). *RMS*, root mean square.

Network	ASM	LM		NN			HNN	
Case	–	2	4	30	50	70	10	70
Parameters	19	3	5	60	100	140	140	980
Training	148	0.1	0.21	3.8	5.28	5.23	2.73	36.81
RMS	37	11.99	11.54	24.39	8.45	10.64	11.81	8.62

age of 15 days. A total of 100 input/output data pairs from the records of 2003 are used as the training data, the other 30 input/output pairs as the testing data. The testing results of effluent COD are shown in Fig. 1.3.

We compare the hierarchical neural networks (HNNs) with the other three modeling methods, i.e., ASMs [4], linear models (LMs) [7], and neural networks (NNs) [6]. The model parameters of the ASM are the default values in [4]. The numbers of concerned variables

in linear models are selected 2, 3, and 4. The hidden nodes of NNs are chosen as 30, 50, and 70; they are the same as those in the HNN. The initial values for all weights are chosen randomly from the interval (0, 1). The comparison results are shown in Table 1.2, where the root mean square (RMS) of the HNN refers to the summation of errors in the final output.

The wastewater treatment process suffers from external disturbances such as temperature, influent qualities, influent flow, operational status, and internal factors like microorganism activities. Both the NN and the HNN can achieve high modeling accuracy. The LM and the NN can provide the water quality in the aerobic reactor, while the ASM and the HNN give water qualities in both anoxic and aerobic reactors.

1.5 CONCLUSIONS

The main contribution of this chapter is that a new hierarchical model is proposed. This hierarchical dynamic neural network is effective for cascade process modeling. Two stable training algorithms are discussed for this model. A new estimation method for the internal variables of the cascade process is illustrated. Real data of a wastewater treatment plant are applied to illustrate the modeling approach.

REFERENCES

1. J. Anderson, H. Kim, T. McAvoy, O. Hao, Control of an alternating aerobic–anoxic activated sludge system — part 1: development of a linearization-based modeling approach, Control Engineering Practice 8 (3) (2000) 271–278, http://www.sciencedirect.com/science/article/pii/S0967066199001744.
2. Q. Cong, W. Yu, T. Chai, Cascade process modeling with mechanism-based hierarchical neural networks, International Journal of Neural Systems 20 (1) (2010) 1–11.
3. A.A. Guergachi, G.G. Patry, Constructing a model hierarchy with background knowledge for structural risk minimization: application to biological treatment of wastewater, IEEE Transactions on Systems, Man, and Cybernetics - Part A: Systems and Humans 36 (2) (March 2006) 373–383.
4. M. Henze, W. Gujer, T. Mino, T. Matsuo, M.C. Wentzel, G.v.R. Marais, M. van Loosdrecht, Activated sludge model no 2d 39 (12 1999) 165–182.
5. L. Jin, M.M. Gupta, Stable dynamic backpropagation learning in recurrent neural networks, IEEE Transactions on Neural Networks 10 (6) (Nov 1999) 1321–1334.
6. D.S. Lee, P.A. Vanrolleghem, J.M. Park, Parallel hybrid modeling methods for a full-scale cokes wastewater treatment plant, Journal of Biotechnology 115 (3) (2005) 317–328, http://www.sciencedirect.com/science/article/pii/S0168165604004742.
7. Z.-C. Lin, W.-J. Wu, Multiple linear regression analysis of the overlay accuracy model, IEEE Transactions on Semiconductor Manufacturing 12 (2) (May 1999) 229–237.
8. M.M. Polycarpou, P.A. Ioannou, Learning and convergence analysis of neural-type structured networks, IEEE Transactions on Neural Networks 3 (1) (Jan 1992) 39–50.
9. G.V.S. Raju, J. Zhou, R.A. Kisner, Hierarchical fuzzy control, International Journal of Control 54 (5) (1991) 1201–1216, https://doi.org/10.1080/00207179108934205.
10. I. Takács, G. Patry, D. Nolasco, A dynamic model of the clarification-thickening process, Water Research 25 (10) (1991) 1263–1271, http://www.sciencedirect.com/science/article/pii/004313549190066Y.
11. L.-X. Wang, A Course in Fuzzy Systems and Control, Prentice-Hall, Inc., Upper Saddle River, NJ, USA, 1997.
12. L.-X. Wang, Analysis and design of hierarchical fuzzy systems, IEEE Transactions on Fuzzy Systems 7 (5) (Oct 1999) 617–624.
13. S. Weijers, Modelling, Identification and Control of Activated Sludge Plants for Nitrogen Removal, Ph.D. dissertation, Technology University of Eindhoven, 2000.
14. W. Yu, X. Li, Discrete-time neuro identification without robust modification, IEE Proceedings - Control Theory and Applications 150 (3) (May 2003) 311–316.
15. W. Yu, F.O. Rodríguez, M.A. Moreno-Armendariz, Hierarchical fuzzy CMAC for nonlinear systems modeling, IEEE Transactions on Fuzzy Systems 16 (5) (Oct 2008) 1302–1314.
16. F.O. Rodríguez, W. Yu, M.A. Moreno-Armendariz, Nonlinear systems identification via two types of recurrent fuzzy CMAC, Neural Processing Letters 28 (1) (2008) 49–62.
17. W. Yu, M.A. Moreno-Armendariz, System identification using hierarchical fuzzy neural networks with stable learning algorithms, in: Proceedings of the 44th IEEE Conference on Decision and Control, Dec 2005, pp. 4089–4094.
18. X. Zeng, J.A. Keane, Approximation capabilities of hierarchical fuzzy systems, IEEE Transactions on Fuzzy Systems 13 (5) (Oct 2005) 659–672.
19. J. Zhu, J. Zurcher, M. Rao, M.Q.-H. Meng, An online wastewater quality predication system based on a time-delay neural network, Engineering Applications of Artificial Intelligence 11 (6) (1998) 747–758, http://www.sciencedirect.com/science/article/pii/S0952197698000177.

CHAPTER 2

Hyperellipsoidal Neural Network Trained With Extended Kalman Filter for Forecasting of Time Series

CARLOS VILLASEÑOR, PHD

2.1 INTRODUCTION

Being able to predict the future of complex phenomena causes a significant impact in industrial applications and scientific research. For example, [24] proposes to use supervised machine learning for diffuse large B cell lymphoma outcome. Forecasting is also essential in investing [7,14] and in projection demand and prices [16].

Nowadays, machine learning has become a primary tool in prediction [20]. In particular, neural networks show excellent capabilities of adaptation and robustness in time series forecasting. Nevertheless, there are still many opportunities and areas to explore that could lead to higher-accuracy algorithms.

In this chapter, a new technique for forecasting of time series using the hyperellipsoidal neural network (HNN) is presented. The HNN is a novel neural network that uses hyperellipsoidal neurons (HNs) [28]. In 2017, the HN was presented. This neuron represents a hyperellipsoid decision surface capable of deforming in other quadratic surfaces, like pseudocylinders and a pair of plains. This neuron is represented in the geometric algebra $\mathcal{G}_{2n,n}$. In [28], the propagation of the neuron is defined with the inner product of the pattern and the ellipsoid, as (2.1) shows, where $E, X \in \mathcal{G}_{2n,2}^1$. We have

$$y = sgn(E \cdot X). \tag{2.1}$$

Although the propagation of the HN is suitable for classification and surface approximation, this chapters presents a modification of the HN propagation using the Mahalanobis distance (MD), which is more suitable for forecasting. The training of the HNN is implemented with the k-means online algorithm [4] for finding the hyperellipsoid center, and the extended Kalman filter (EKF) is used for training the hyperellipsoid semiaxes and the output layer.

The EKF performance depends on the chosen covariance matrices. Then, the proposal uses the germinal center optimization (GCO) [29] algorithm to optimize the covariance parameters of the EKF. GCO is an artificial immune system inspired by the high-affinity antibodies produced in the germinal center reaction. We conducted six experiments to demonstrate the proposal capabilities, including a comparison with the ADALINE algorithm and the results of Fuzzy type-2 proposals published in [25].

2.2 MATHEMATICAL BACKGROUND

To clarify the proposed scheme, in this section we include a brief review of the mathematical tools useful for the HNN propagation and training.

2.2.1 Mahalanobis Distance

The MD [19,8] is a distance measure that takes into account the correlation in the data by using the precision matrix (inverse of the covariance matrix). Consider the covariance described in (2.2), where E denotes the expected value of a probability distribution. We have

$$\sigma(x, y) = E[(x - E[x])(y - E[y])]$$

$$= \frac{1}{n} \sum_{i=1}^{n} (x_i - \bar{x})(y_i - \bar{y}). \tag{2.2}$$

Then for a vector $x = [x_1, x_2, \ldots, x_n]^T \in \mathcal{R}^n$ we could calculate the covariance matrix S with (2.3); S is closely related to the correlation matrix C and we can find the matrix of the Pearson product–moment correlation coefficient with

$$S = \begin{bmatrix} \sigma(x_1, x_1) & \sigma(x_1, x_2) & \cdots & \sigma(x_1, x_n) \\ \sigma(x_2, x_1) & \sigma(x_2, x_2) & \cdots & \sigma(x_2, x_n) \\ \vdots & \vdots & \ddots & \vdots \\ \sigma(x_n, x_1) & \sigma(x_n, x_2) & \cdots & \sigma(x_n, x_n) \end{bmatrix}, \tag{2.3}$$

$$C = (\text{diag}(S))^{-\frac{1}{2}} S (\text{diag}(S))^{-\frac{1}{2}}. \tag{2.4}$$

Artificial Neural Networks for Engineering Applications. https://doi.org/10.1016/B978-0-12-818247-5.00011-3

9

With these definitions we are able to calculate the MD for two vectors $x, y \in \mathcal{R}^n$ with (2.5), where S^{-1} is the precision matrix. The MD satisfies the following properties:

$$d_M(x, y) = \sqrt{(x - y)^T S^{-1}(x - y)}, \qquad (2.5)$$

$$d_M(x, y) = d_M(y, x), \qquad (2.6)$$

$$d_M(x, y) = 0 \leftrightarrow x = y, \qquad (2.7)$$

$$d_M(x, y) \leq d_M(x, z) + d_M(z, y). \qquad (2.8)$$

Consider the eigenvalues $\{\lambda_i\}_{i=1}^m \subset \mathcal{R}$ and the eigenvectors $\{e_i\}_{i=1}^m \subset \mathcal{R}^n$ of the covariance matrix S. Then we could apply spectral decomposition of the following precision matrix:

$$S^{-1} = \sum_{i=1}^n \frac{1}{\lambda_i} e_i e_i^T. \qquad (2.9)$$

The covariance ellipsoid is a geometric interpretation of the covariance. It is formed using the mean as the center of the ellipsoid, the eigenvalues as the semiaxes, and the eigenvector as the orientation. In Fig. 2.1, we present an example of the covariance ellipsoid.

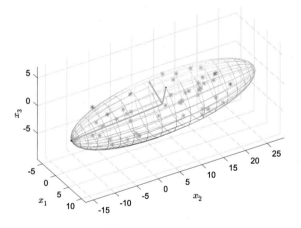

FIG. 2.1 Covariance ellipsoid representation.

The MD distance represents a statistical measure in a joint probability distribution. Nowadays, MD is a popular tool for many applications like clustering and classification [32], principal component analysis and pattern recognition [21], biometric information treatment [23], and econometric applications [26].

In this chapter, we use the MD for operating every input pattern in the HNN, where every HN represents a covariance ellipsoid.

2.2.2 Extended Kalman Filter

The Kalman filter [5] is a linear quadratic stochastic estimator of unknown variables under Gaussian noise. We can use the Taylor expansion to make the extended Kalman filter suitable for nonlinear problems.

Consider the system in (2.10) and (2.11), where f and h are two differentiable functions, x is the state vector, u is the input vector, and q_k and r_k are the process and observation noises, respectively, which are assumed to be Gaussian with zero mean and covariance Q and R, respectively. We have

$$x_k = f(x_{k-1}, u_k) + q_k, \qquad (2.10)$$

$$z = h(x_k) + r_k. \qquad (2.11)$$

Then the EKF [10] is defined with a predict step in (2.12) and (2.13), and a correct step in (2.14)–(2.16), where $\hat{x}_{k|k-1}$ is the predicted $\hat{x}_k$ due to the information in the time $k - 1$, $P_{k|k-1}$ is the predicted covariance estimate, and K is the Kalman gain. We have

$$\hat{x}_{k|k-1} = f(\hat{x}_{k-1|k-1}, u_k), \qquad (2.12)$$

$$P_{k|k-1} = F_k P_{k-1|k-1} F_k^T + Q_k, \qquad (2.13)$$

$$K_k = P_k H_k^T \left[R_k + H_k P_k H_k^T \right]^{-1}, \qquad (2.14)$$

$$\hat{x}_{k|k} = \hat{x}_{k|k-1} + K_k(z_k - h(\hat{x}_{k|k-1})), \qquad (2.15)$$

$$P_{k|k} = (I - K_k H_k) P_{k|k+1}. \qquad (2.16)$$

The terms F_k and H_k, defined in (2.17) and (2.18), respectively, are Jacobians of the f and h functions of the system. We have

$$F_k = \left. \frac{\partial f}{\partial x} \right|_{\hat{x}_{k-1|k-1}, u_k} \qquad (2.17)$$

$$H_k = \left. \frac{\partial h}{\partial x} \right|_{\hat{x}_{k|k-1}} \qquad (2.18)$$

The EKF has become a favorite tool in nonlinear state estimation, for example in navigation systems and GPS [30], traffic estimation [31], mapping and localization [13], robot dynamic identification [11], and neurocontrol [2]. In this chapter, we use the EKF for training the semiaxes of the hyperellipsoids in the HNN and the weights in the output layer.

2.2.3 K-Means Online

K-means [18] is an algorithm for vector quantization by partitional clustering. Consider a set of observations $\{x_i\}_{i=1}^d \subset \mathcal{R}^n$. Then the k-means aims to find a partition, in the sense of Voronoi [3], of the n observations into k sets denoted by $A = \{A_i\}_{i=1}^k$ such that (2.19) is minimized, where μ_i is the mean of the cluster A_i. The standard implementation of the k-means problem is an algorithm of Lloyd [17]. We have

$$\arg\min_A \sum_{i=1}^k \sum_{x \in A_i} ||x - \mu_i||^2. \qquad (2.19)$$

This algorithm contains an assignation step (defined in (2.20)), where every pattern x_p is assigned to exact one cluster A_i, and the update state in (2.21), where the new centroids are calculated. These two steps are applied iteratively until no pattern is reassigned. We have

$$A_i = \{x_p : ||x_p - \mu_i||^2 \le ||x_p - \mu_j||, \forall j \in \{1, 2, \ldots, k\}\}, \qquad (2.20)$$

$$\mu_i = \frac{1}{|A_i|} \sum_{x \in A_i} x. \qquad (2.21)$$

However, this standard implementation of the k-means is not available to get online data as the forecasting problem requires. For this reason, the proposed HNN for forecasting uses an online version presented in [4] and described in Algorithm 1 to train the center of the hyperellipsoids.

Algorithm 1 K-means online.

Data: Pattern x_i, Current centroids $\{\mu_t\}_{t=1}^k$, Learning rate ζ
Result: Updated centroids $\{\mu_t\}_{t=1}^k$
1 **if** $i \le k$ **then**
2 $\quad$ $\mu_i \leftarrow x_i$
3 **else**
4 $\quad$ min $\leftarrow$ MAX_DBL ;   // MAX_DBL: maximum in double
$\qquad$ data type
5 $\quad$ **for** $j \in \{1, 2, \ldots, k\}$ **do**
6 $\qquad$ **if** $||x_i - \mu_j|| < min$ **then**
7 $\qquad\quad$ $l \leftarrow j$ min $\leftarrow ||x_i - \mu_j||$
8 $\qquad$ **end**
9 $\quad$ **end**
10 $\quad$ $\mu_l \leftarrow \mu_l + \zeta(x_i - \mu_l)$
11 **end**

2.2.4 Germinal Center Optimization

In this section, we briefly review the GCO algorithm [29]. GCO is a metaheuristic for multivariate continuous optimization based on swarm intelligence [22,15] and the vertebrate adaptive immune system [1]. The GCO algorithm is an artificial immune system [33]

but with some methodologies of evolutionary computing [9].

The vertebrate immune system is the set of biological mechanisms with which the body can protect itself. It is made up of two subsystems, the innate immunity and the adaptive immunity. The innate immunity is a nonspecific defense that takes place in the first hours of the appearance of antigens (Ag) (foraging substances). It is formed of physical barriers (such as skin), mast cells, macrophages, phagocytes, and neutrophils. The innate immunity has the task of the immediate defense against infection and is an evolutionary trait.

However, the innate immunity is not always capable of defending the body. Under these circumstances a more specific immune response is needed it and this is provided by the adaptive immunity. The adaptive immunity is a complex mechanism formed mainly by B lymphocytes (B cells) and T lymphocytes (T cells). B cells are capable of differentiating into plasma cells to release antibodies (Ab). The Ab attach to the epitope of the Ag to enable another immune cell to phagocytize it. The Ab has a hypervariable area to attach to the Ab; the force of the attachment is called affinity.

The B cells compete with each other to get the best affinity, rewarded by the helper T cells. This competition enables the body to be protected against specific Ag. The process by which the highest Ab affinity is reached is called germinal center reaction [27]. The germinal centers are sites in the secondary lymphoid tissues. Here the B cells get into an iterative affinity refining process. The GC reaction comprises two histologically recognizable zones. The first is the dark zone, where clonal expansion and somatic hypermutation (SHM) take place. The dark zone aims to diversify the B cells.

The second is the light zone, where there is a competition for antigen reserves among the follicular dendritic cells (FDCs) and competition to get a life signal from helper T cells takes place. This life signal allows the B cells to live longer and proliferate more. Afterwards, the B cell can reenter the dark zone, but the probability it could get out of the GC and differentiate into a plasma cell or a memory B cell is low. In Fig. 2.2, we show a simplification of the GC reaction.

Inspired by this topic and by the classical metaheuristic algorithms like differential evolution (DE) [22] and particle swarm optimization (PSO) [15], GCO implements the dark zone and light zone processes, where the B cells and Ab represent candidate solutions and the T cells represent a reward system. The nonuniform probability distribution of the B cell multiplicity imitates clonal expansion. In the same way as DE, this algorithm chooses three candidate solutions to mutate

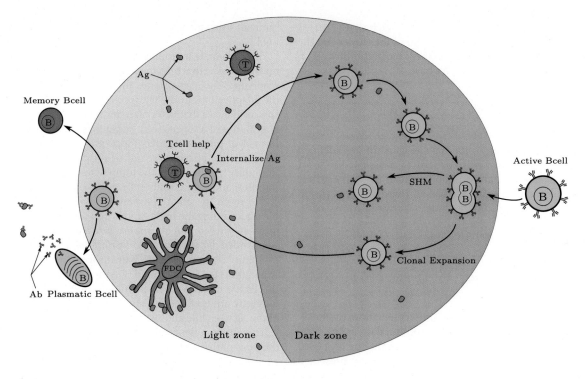

FIG. 2.2 Germinal center reaction.

to the current solution, but unlike DE these solutions are chosen through the probability distribution. Then, this distribution is modeled according to the problems, and the leadership adapts to the fitness function.

In Fig. 2.3, we present the GCO algorithm flowchart.

2.3 HNN FOR TIME SERIES FORECASTING

In this section, we present the proposed scheme for forecasting with HNN. Consider Fig. 2.4. Let us use $u \in \mathcal{R}^{p+1}$, the input vector defined in (2.22), and the hyperellipsoid with center in $c = [c_1, c_2, \ldots, c_{p+1}]^T$ and semiaxes $\{r_i\}_{i=1}^{p+1}$. If this representation of the neuron is fixed in the coordinate system, then no rotation is assumed. Thus, the eigenvectors of the covariance ellipsoid are the base of the canonical base of the Euclidean space. We have

$$u(k) = \left[f(x(k)), f(x(k-1)), \ldots, f(x(k-p)) \right]^T, \tag{2.22}$$

$$\frac{(x_1 - c_1)^2}{r_1^2} + \frac{(x_2 - c_2)^2}{r_2^2} + \cdots + \frac{(x_{p+1} - c_{p+1})^2}{r_{p+1}^2} = 0. \tag{2.23}$$

Then, the propagation of every HN is given by (2.24), where S^{-1} is defined in (2.25). We have

$$y_i = \sqrt{(u - c)^T S^{-1} (u - c)}, \tag{2.24}$$

$$S^{-1} = \begin{bmatrix} \frac{1}{r_1} & 0 & \cdots & 0 \\ 0 & \frac{1}{r_2} & \cdots & 0 \\ \vdots & \vdots & \ddots & \vdots \\ 0 & 0 & \cdots & \frac{1}{r_{p+1}} \end{bmatrix}. \tag{2.25}$$

Finally, the HNN propagation function is defined in (2.26). The center is trained with k-means online. For the other parameters consider the state vector W in (2.27), where $r_{i,j}$ is the j semiaxis of the HN$_i$. To train the state vector W with EKF consider $F = I$ and the function h as the HNN propagation. We have

$$\hat{f}(x(k+1)) = h(u)$$
$$= \sum_{i=1}^{p+1} w_i \sqrt{(u - c_i)^T S_i^{-1} (u - c_i)} + w_0, \tag{2.26}$$

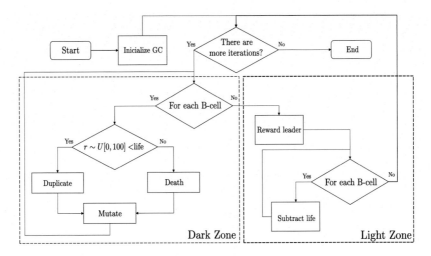

FIG. 2.3 GCO algorithm flowchart.

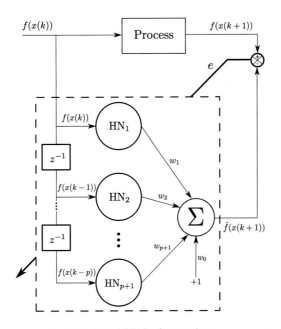

FIG. 2.4 HNN for forecasting.

$$W = [r_{1,1}, \ldots, r_{1,p+1}, \ldots, r_{p+1,1}, \ldots, r_{p+1,p+1},$$
$$w_1, w_2, \ldots, w_{p+1}, w_0]^T.$$
$$(2.27)$$

Then we use backpropagation to calculate the local gradients in (2.28), where $e = f(x(k+1)) - \hat{f}(x(k+1))$ and $c_{i,j}$ is the j coordinate of the center vector in the HN_i. Then, the Jacobian H of the EKF is defined in (2.29) and the training is similar to that of a multilayer perceptron [6], i.e.,

$$\frac{\partial h(u)}{\partial r_{i,j}} = e w_i \left(\frac{u_i - c_{i,j}}{r_{i,j}} \right)^2 = e \gamma_{i,j}, \qquad (2.28)$$

$$H = [\gamma_{1,1}, \ldots, \gamma_{1,p+1}, \ldots, \gamma_{p+1,1}, \ldots, \gamma_{p+1,p+1},$$
$$y_1, \ldots, y_{p+1}, 1]^T. \qquad (2.29)$$

2.4 RESULTS

In this section, we show six experiments that demonstrate the capability of the proposed scheme. The matrices P, Q, and R of the EKF are considered to be diagonals of the form kI, where $k \in \mathcal{R}$ and I is the identity matrix. The GCO optimizes these values for each experiment; the resulting values are shown in Table 2.1. In the same table the number of iterations is reported, which represents the number of times that we trained with the same pattern in an epoch. For all the six experiment we use a regression vector of size 2 and seven HNs.

2.4.1 Comparison With the ADALINE Algorithm

In this experiment, we compare the proposed scheme with the ADALINE algorithm, in order to present the advantage of the HNs. In Fig. 2.5, we show the results of the ADALINE for the following series:

$$f(x) = 5\sin(2x) + 2\sin(\tfrac{1}{2}x) + 4\sin(x) + 6 + \text{noise}.$$
$$(2.30)$$

TABLE 2.1
EKF optimal values found by GCO and performed iterations.

Experiment	P	Q	R	Iterations
1	127.5701	44.1933	760.6742	5
2	1507.600	45.6573	442470	25
3	479.3557	4.9616	1372000	16
4	2089.34	13.5572	38.6392	3
5	449649	1.1477	16950	25
6	703.9904	923.284	619.1079	5

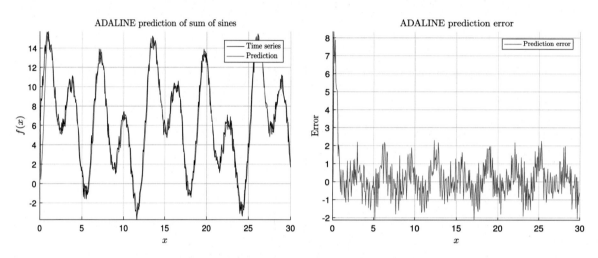

FIG. 2.5 Experiment 1. Optimal ADALINE for sum of sines time series.

The ADALINE in Fig. 2.5 was trained with EKF with optimal values $P = 251.469$, $Q = 32.246$, and $R = 568.424$. Moreover, five iterations were applied for prediction. In Fig. 2.6, we present the HNN forecasting for the same time series. In Table 2.2, we can find the numerical comparison of the algorithms, where it is easy to see the superior accuracy of the proposed scheme.

2.4.2 Experiments With Real-Time Series

The following four experiments are forecasting of real-time series, all of them from DataMarket of Qlink enlisted below. The results are shown in Figs. 2.7–2.10 and in Table 2.3.

- Experiment 2: Daily minimum temperatures in Melbourne, Australia, 1981–1990.
- Experiment 3: Zürich monthly sunspot numbers 1749–1983.
- Experiment 4: Euro/ECU exchange rates – monthly data (Mexican peso).

- Experiment 5: Daily total female births in California, 1959.

The results indicate an excellent performance of the HNN, even with high-frequency data.

2.4.3 Mackey–Glass Equation

For the last experiment, we present the forecasting of the Mackey–Glass equation [12], which has become a benchmark for forecasting of time series. The Mackey–Glass equation is a nonlinear time delay differential equation defined by (2.31). In Fig. 2.11 and Table 2.4, we show the results of the Mackey–Glass forecasting. We have

$$\frac{dx}{dt} = \beta \frac{x(t - \tau)}{1 - x(t - \tau)^n} - \gamma x \quad \gamma, \beta, n > 0. \quad (2.31)$$

Other approaches, for example the one from [25], which uses an interval fuzzy type-2 integrator, obtain in the best case (with a Gauss membership function)

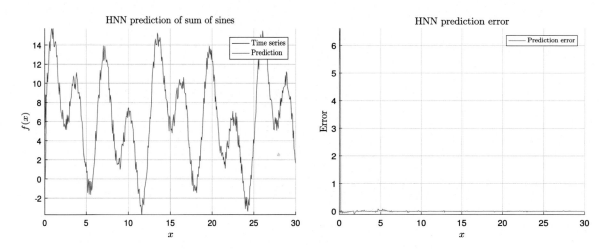

FIG. 2.6 Experiment 1. HNN for sum of sines time series.

TABLE 2.2
Results of Experiment 1.

Algorithm	Accumulated error	Mean error	STD	RMSE
ADALINE	394.7103	0.7894	0.9103	0.7251
HNN	22.1798	0.0444	0.4698	0.1111

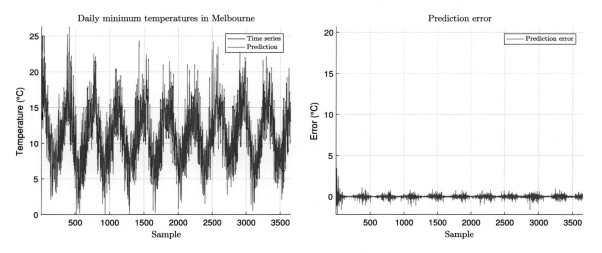

FIG. 2.7 Experiment 2.

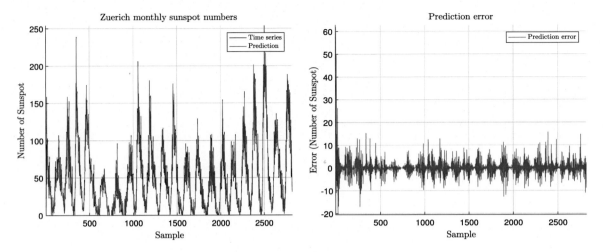

FIG. 2.8 Experiment 3.

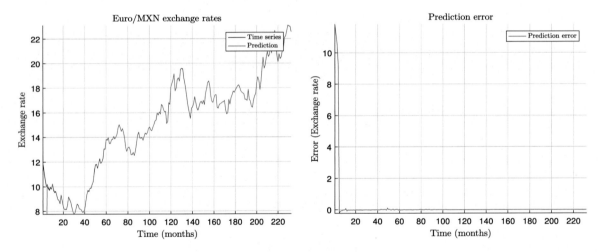

FIG. 2.9 Experiment 4.

TABLE 2.3 Results of Experiments 2 to 5.				
Experiment	**Accumulated error**	**Mean error**	**STD**	**RMSE**
2	489.372	0.1341	0.4866	0.1273
3	5425.2	1.9238	3.3319	7.3992
4	45.2237	0.1949	1.4038	1.0001
5	148.8096	0.4077	1.859	1.8064

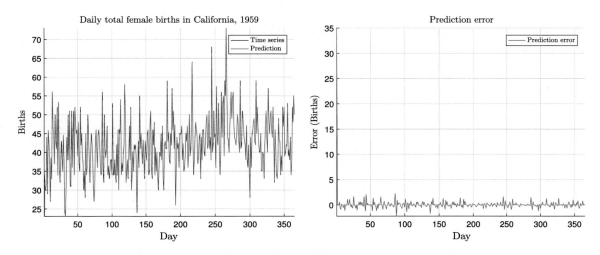

FIG. 2.10 Experiment 5.

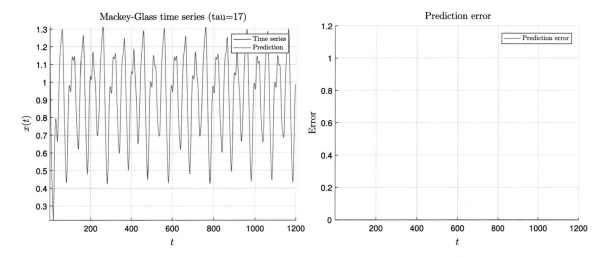

FIG. 2.11 Experiment 6.

TABLE 2.4
Results of Experiment 6.

Experiment	Accumulated error	Mean error	STD	RMSE
6	3.4194	0.02849	0.01675	1.4038E−04

an RMSE $= 0.011248$, which is greater than the HNN RMSE $= 1.4038E-04$. Therefore, our proposed scheme shows a very high accuracy.

2.5 CONCLUSION

In the present chapter, we develop a novel technique for time series forecasting. We defined the HNN based on the covariance ellipsoid and deduced its propagation with the Mahalanobis distance. We show hybrid training capable of online learning, using k-means online for training the center of the ellipsoids and EKF for training the semiaxes and the weights of the output layer. The EKF covariances were optimized with the GCO algorithm.

We demonstrated the differences between the HNN and the optimized ADALINE. The proposed algorithm shows high accuracy results in public data sets and better accuracy in comparison to the interval fuzzy type-2 approach in the Mackey–Glass equation.

Finally, the HNN shows high performance in forecasting of time series without needing recursion connections.

REFERENCES

1. A.K. Abbas, A.H. Lichtman, S. Pillai, Cellular and Molecular Immunology e-Book, Elsevier Health Sciences, 2014.
2. A.Y. Alanis, J.D. Rios, J. Rivera, N. Arana-Daniel, C. Lopez-Franco, Real-time discrete neural control applied to a linear induction motor, Neurocomputing 164 (2015) 240–251.
3. F. Aurenhammer, Voronoi diagrams—a survey of a fundamental geometric data structure 23 (3) (1991) 345–405.
4. W. Barbakh, C. Fyfe, Online clustering algorithms, International Journal of Neural Systems 18 (03) (2008) 185–194.
5. G. Bishop, G. Welch, et al., An introduction to the Kalman filter, in: Proc. of SIGGRAPH, Course 8, 2001, p. 41.
6. E.N.S. Camperos, A.Y.A. García, Redes neuronales: conceptos fundamentales y aplicaciones a control automático, Pearson Educación, 2006.
7. A. Cowles, Stock market forecasting, Econometrica, Journal of the Econometric Society (1944) 206–214.
8. R. De Maesschalck, D. Jouan-Rimbaud, D.L. Massart, The Mahalanobis distance, Chemometrics and Intelligent Laboratory Systems 50 (1) (2000) 1–18.
9. A.E. Eiben, J.E. Smith, et al., Introduction to Evolutionary Computing, vol. 53, Springer, 2003.
10. K. Fujii, Extended Kalman Filter, Reference Manual, 2013.
11. M. Gautier, P. Poignet, Extended Kalman filtering and weighted least squares dynamic identification of robot, Control Engineering Practice 9 (12) (2001) 1361–1372.
12. L. Glass, M. Mackey, Mackey–Glass equation, Scholarpedia 5 (3) (2010) 6908.
13. G.P. Huang, A.I. Mourikis, S.I. Roumeliotis, Analysis and improvement of the consistency of extended Kalman filter based SLAM, in: Robotics and Automation, 2008, ICRA 2008, IEEE International Conference on, IEEE, 2008, pp. 473–479.
14. W. Huang, Y. Nakamori, S.-Y. Wang, Forecasting stock market movement direction with support vector machine, Computers & Operations Research 32 (10) (2005) 2513–2522.
15. J. Kennedy, Particle swarm optimization, in: Encyclopedia of Machine Learning, Springer, 2011, pp. 760–766.
16. G. Li, H. Song, S.F. Witt, Recent developments in econometric modeling and forecasting, Journal of Travel Research 44 (1) (2005) 82–99.
17. S. Lloyd, Least squares quantization in PCM, IEEE Transactions on Information Theory 28 (2) (1982) 129–137.
18. J. MacQueen, et al., Some methods for classification and analysis of multivariate observations, in: Proceedings of the Fifth Berkeley Symposium on Mathematical Statistics and Probability, vol. 1, no. 14, Oakland, CA, USA, 1967, pp. 281–297.
19. P.C. Mahalanobis, On the generalized distance in statistics, Proceedings of National Institute of Science of India 2 (1936) 49–55.
20. D.P. Mandic, J.A. Chambers, et al., Recurrent Neural Networks for Prediction: Learning Algorithms, Architectures and Stability, Wiley Online Library, 2001.
21. V. Perlibakas, Distance measures for PCA-based face recognition, Pattern Recognition Letters 25 (6) (2004) 711–724.
22. K. Price, R.M. Storn, J.A. Lampinen, Differential Evolution: A Practical Approach to Global Optimization, Springer Science & Business Media, 2006.
23. A. Ross, A. Jain, Information fusion in biometrics, Pattern Recognition Letters 24 (13) (2003) 2115–2125.
24. M.A. Shipp, K.N. Ross, P. Tamayo, A.P. Weng, J.L. Kutok, R.C. Aguiar, M. Gaasenbeek, M. Angelo, M. Reich, G.S. Pinkus, et al., Diffuse large b-cell lymphoma outcome prediction by gene-expression profiling and supervised machine learning, Nature Medicine 8 (1) (2002) 68.
25. J. Soto, P. Melin, O. Castillo, Optimization of interval type-2 fuzzy integrators in ensembles of ANFIS models for prediction of the Mackey–Glass time series, in: Norbert Wiener in the 21st Century (21CW), 2014 IEEE Conference on, IEEE, 2014, pp. 1–8.
26. A. Ullah, Entropy, divergence and distance measures with econometric applications, Journal of Statistical Planning and Inference 49 (1) (1996) 137–162.
27. G.D. Victora, M.C. Nussenzweig, Germinal centers, Annual Review of Immunology 30 (2012) 429–457.
28. C. Villaseñor, N. Arana-Daniel, A.Y. Alanis, C. Lopez-Franco, Hyperellipsoidal neuron, in: Neural Networks (IJCNN), 2017 International Joint Conference on, IEEE, 2017, pp. 788–794.
29. C. Villaseñor, J.D. Rios, N. Arana-Daniel, A.Y. Alanis, C. Lopez-Franco, E.A. Hernandez-Vargas, Germinal center optimization applied to neural inverse optimal control for an all-terrain tracked robot, Applied Sciences 8 (1) (2017) 31.

30. E. Wan, Sigma-point filters: an overview with applications to integrated navigation and vision assisted control, in: Nonlinear Statistical Signal Processing Workshop, 2006 IEEE, IEEE, 2006, pp. 201–202.

31. Y. Wang, M. Papageorgiou, Real-time freeway traffic state estimation based on extended Kalman filter: a general approach, Transportation Research Part B: Methodological 39 (2) (2005) 141–167.

32. S. Xiang, F. Nie, C. Zhang, Learning a Mahalanobis distance metric for data clustering and classification, Pattern Recognition 41 (12) (2008) 3600–3612.

33. B.-H. Yang, Introduction to a novel optimization method: artificial immune systems, IE Interfaces 20 (4) (2007) 458–468.

CHAPTER 3

Neural Networks: A Methodology for Modeling and Control Design of Dynamical Systems

FERNANDO ORNELAS-TELLEZ, PHD • J. JESUS RICO-MELGOZA, PHD •
ANGEL E. VILLAFUERTE, PHD • FEBE J. ZAVALA-MENDOZA, MSC

3.1 INTRODUCTION

Nonlinear systems usually present complex and often unpredictable behaviors in natural phenomena. Nonlinear systems are present in different areas of everyday life, e.g., engineering, industrial processes, economic data, biology, life sciences, medicine, and health care. It is difficult to have general modeling methods due to the variety of nonlinear systems, which are usually subject to uncertainties and disturbances. Therefore, it is convenient to use alternative methodologies for modeling, such as identification schemes, i.e., the estimation of dynamic system models from observed data, among other methodologies. Once an adequate system identification is achieved, the resulting model can be used for various purposes such as control design, state variable estimation, and prediction.

For linear systems, there exist well-structured theories, methodologies, and algorithms for their respective modeling and control. For the case of nonlinear systems, the situation is more complex. In this sense, the identification method is only as good as the model it utilizes, depending on of the input–output relations of the system that can be measured. In the literature different methods are described, which are used to identify nonlinear systems [50,48], such as adaptive schemes [4,36], polynomial identifiers [45,23], and neural and polynomial networks [49]. These identifier methods propose nonlinear models whose structures are based on parameters, named neural weights, to be determined or adapted.

Two schemes are common in the field of system identification: (a) a black-box methodology, where only the data of the relationships between the input and output are known, which should serve to determine an identifier model, and (b) a gray-box methodology, where a priori knowledge of the system (e.g., system order, structure, variables relationships) can be used to propose or determine an identifier model. The last

methodology could be more effective to determine accurate models because there exists previous knowledge of the nonlinear system to be studied, allowing to incorporate a priori information and characteristics about the system. Accordingly, a viable alternative to model unknown nonlinear systems is the usage of neural networks (NNs) to identify their dynamics. In addition, for control purposes the NN models should be easy to implement, have a relatively simple structure, be robust against disturbances and parameter changes, and have the capacity to adjust their parameters (weights) online [47,28,57,61]. One of the well-known learning algorithms for NNs is the backpropagation training method; however, it is based on a gradient descent technique and hence its learning speed is usually slow [63]. Recently, extended Kalman filter (EKF)-based algorithms have been introduced to train NNs, with the aim of improving the learning convergence [63]. References [55,57,59] have reviewed the application of recurrent high-order NNs (RHONNs) for nonlinear system neural modeling and control design. In [59], an adaptive neural identification and control scheme through online learning is analyzed, where the closed-loop system stability is stated using the Lyapunov function method, where the neural learning is performed online through an EKF [59,28]. Another interesting alternative are the polynomial NNs (PNN), a flexible architecture whose structure could be developed through learning [51]. Among the main features of PNNs, one can find that the polynomials have adequate capabilities to approximate nonlinear functions and that easy interrelationships between the significant input variables can be established, as well as exploitation of different orders of the used polynomials (say, linear, quadratic, cubic, etc.).

Both a RHONN and a PNN are proposed in this chapter, which are used for modeling and control purposes. Firstly, a discrete-time recurrent NN (RNN) de-

Artificial Neural Networks for Engineering Applications. https://doi.org/10.1016/B978-0-12-818247-5.00012-5

sign is presented for the modeling of unknown dynamics in nonlinear systems; once the neural identifier is stated, a controller is synthesized based on the discrete-time block control technique combined with sliding modes, which consider constrained inputs. Secondly, a continuous-time adaptive PNN and a nonlinear robust optimal tracking control methodology are developed to model and control uncertain nonlinear systems, where the PNN weights are adjusted online using a Kalman–Bucy filter, an efficient algorithm for neural learning. Engineering examples are used to illustrate the applicability of the NN-based methodology. Although two control methodologies are developed in this chapter, other neural controllers could be proposed based on adequate NN designs. The organization of the chapter is as follows. Section 3.2 and Section 3.3 present the neural modeling control designs for discrete-time and continuous-time systems, respectively. Finally, other NN usages are briefly described in Section 3.4, and general conclusions are given in Section 3.5.

This chapter presents the following main salient features and flexibilities of NN for modeling and control purposes in dynamical systems:

- The NN can be designed with a convenient structure for control purposes. In Section 3.2.3, a RHONN is stated in the controllable canonical form such that the block control technique is used. In Section 3.3.2 a PNN is designed with the aim of synthesizing a nonlinear optimal control to achieve trajectory tracking.
- Under the assumption that a real system is controllable, the controllability property must be guaranteed in the NN designs (see Remark 3.1 and Remark 3.8). The observability property can be ensured in a similar way.
- NNs can be used for dealing with large-scale systems by using a decentralized approach, where large systems can be decomposed into low-order interconnected subsystems for easy analysis and control purposes. The subsystem interconnections' terms are modeled by disturbances, being the NNs robust against those terms (Section 3.2.4.2).

3.2 NEURAL MODELING AND CONTROL FOR DISCRETE-TIME SYSTEMS

Analysis and control of uncertain nonlinear systems is a difficult task, since the real parameters are usually hard to obtain [25]. For real applications, a control scheme based on a system model could not perform as desired, due to external and internal disturbances, or unmodeled dynamics [25]. This fact motivates the

research community in deriving artificial models based on RHONNs to identify the dynamics of uncertain systems.

This section analyzes a general class of nonlinear systems, with a disturbance as studied in [14]; and hence, a similar structure is considered for the neural identifier.

3.2.1 Discrete-Time Uncertain Nonlinear Systems

Let us consider a class of discrete-time disturbed nonlinear system,

$$\chi_{k+1} = f(\chi_k) + g(\chi_k)u_k + \Gamma_k, \qquad (3.1)$$

where $\chi_k \in \mathbb{R}^n$ is the system state at time k and $\Gamma_k \in \mathbb{R}^n$ is an unknown and bounded perturbation term representing uncertain parameters, disturbances, and modeling errors; $f : \mathbb{R}^n \to \mathbb{R}^n$ and $g : \mathbb{R}^n \to \mathbb{R}^{n \times m}$ are smooth maps. Without loss of generality, it is assumed that $\chi_k = 0$ is a fixed point for (3.1). Assume $f(0) = 0$ and $rank\{g(\chi_k)\} = m \ \forall \chi_k \neq 0$. Let us consider that system (3.1) is controllable and observable.

3.2.2 Discrete-Time Recurrent High-Order Neural Network

For the identification of system (3.1), let us consider the discrete-time RHONN proposed in [59], as follows:

$$x_{i,k+1} = w_{i,k}^T \rho_i(x_k, u_k), \qquad (3.2)$$

where $x_k = [x_{1,k} \ x_{2,k} \ \cdots \ x_{n,k}]^T$, x_i is the state of the ith neuron, with $i = 1, \ldots, n$, where x_i identifies the ith component of the state variable χ_k in (3.1); the term w_i is the corresponding online adapted weight vector, and $u_k = [u_{1,k} \ u_{2,k} \ \cdots \ u_{m,k}]^T$ is the neural network input; ρ_i is an L_p-dimensional vector defined as

$$\rho_i(x_k, u_k) = \begin{bmatrix} \rho_{i_1} \\ \rho_{i_2} \\ \vdots \\ \rho_{i_{L_p}} \end{bmatrix} = \begin{bmatrix} \prod_{\ell \in I_1} z_{i_\ell}^{d_{i_\ell}(1)} \\ \prod_{\ell \in I_2} z_{i_\ell}^{d_{i_\ell}(2)} \\ \vdots \\ \prod_{\ell \in I_{L_p}} z_{i_\ell}^{d_{i_\ell}(L_p)} \end{bmatrix},$$

$$(3.3)$$

where L_p is the respective number p of high-order connections, d_{i_ℓ} are nonnegative integers, and $\{I_1, I_2, \ldots, I_{L_p}\}$ is a collection of nonordered subsets of $\{1, 2, \ldots,$

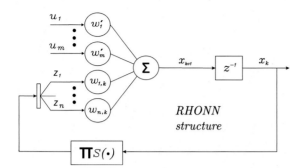

FIG. 3.1 RHONN block diagram.

$n + m$}. Vector Z_i is defined as

$$Z_i = \begin{bmatrix} Z_{i_1} \\ \vdots \\ Z_{i_n} \\ Z_{i_{n+1}} \\ \vdots \\ Z_{i_{n+m}} \end{bmatrix} = \begin{bmatrix} S(x_{1,k}) \\ \vdots \\ S(x_{n,k}) \\ u_{1,k} \\ \vdots \\ u_{m,k} \end{bmatrix},$$

where function $S(\cdot)$ is defined by

$$S(x) = \frac{\alpha_i}{1 + e^{-\beta_i x}} - \gamma_i \qquad (3.4)$$

with $S(\cdot) \in [-\gamma_i, \alpha_i - \gamma_i]$; α_i, β_i, and γ_i are positive constants. A RHONN block diagram is depicted in Fig. 3.1.

The following remark states an important property that the discrete-time RHONN (3.2) structure should consider.

Remark 3.1. The neural weights related to the control inputs can be selected fixed (w_i') to ensure identifier controllability.

Based on Remark 3.1 and using the structure of system (3.1), the following NN model to identify (3.1) is proposed as

$$x_{i,k+1} = w_{i,k}^T \rho_i(x_k) + w_i'^T \psi_i(x_k, u_k), \qquad (3.5)$$

where x_i is the ith neuron state; w_i' is the fixed weight vector and $w_{i,k}$ is the online adaptable weight vector; ψ denotes a function of x or u accordingly to the plant structure (3.1) or external inputs of the identifier, respectively. Vector ρ_i in (3.5) is as defined in (3.3); however, Z_i becomes

$$Z_i = \begin{bmatrix} Z_{i_1} \\ \vdots \\ Z_{i_n} \end{bmatrix} = \begin{bmatrix} S(x_{1,k}) \\ \vdots \\ S(x_{n,k}) \end{bmatrix}.$$

The online adaptable weight vector $w_{i,k}$ is given as

$$w_{i,k}^T = \begin{bmatrix} w_{i\,1,k} & \cdots & w_{i\,L_p,k} \end{bmatrix}.$$

Remark 3.2. It is worth noting that (3.5) does not consider the disturbance term (Γ_k) because the RHONN weights are adapted online, and thus the RHONN models the nonlinear system dynamics, absorbing the disturbance effects.

3.2.2.1 RHONN Models

As described in [57], a RHONN to model a system, as the one given in (3.1), exists; thereby, system (3.1) can be described by the neural identifier

$$\chi_{k+1} = W_k^* \rho(\chi_k) + W'^* \psi(\chi_k, u_k) + v_k, \qquad (3.6)$$

where $W_k^* = [w_{1,k}^{*T} \ w_{2,k}^{*T} \dots w_{n,k}^{*T}]^T$ and $W'^* = [w_1'^{*T} \ w_2'^{*T} \dots w_n'^{*T}]^T$ are the optimal unknown weight matrices, while the modeling error v_k can be calculated as

$$v_k = f(\chi_k) + g(\chi_k) u_k + \Gamma_k - W_k^* \rho(\chi_k) - W'^* \psi(\chi_k, u_k).$$

The modeling error v_k can be made arbitrarily small by appropriately selecting L_p, that is, the number of high-order neural interconnections [57]. The ideal weight matrices W_k^* and W'^* are artificial quantities required for purposes of the stability and convergence analysis. In general, it is assumed the ideal weights exist and are constant but unknown, hence $w_{i,k}^*$ is approximate by the online adjustable ones $w_{i,k}$ [59].

For neural identification of (3.1), two possible models for (3.5) can be used:
- the parallel model

$$x_{i,k+1} = w_{i,k}^T \rho_i(x_k) + w_i'^T \psi_i(x_k, u_k); \qquad (3.7)$$

- the series-parallel model

$$x_{i,k+1} = w_{i,k}^T \rho_i(\chi_k) + w_i'^T \psi_i(\chi_k, u_k). \qquad (3.8)$$

3.2.2.2 Online Learning Law

For the RHONN weights adaptation, an EKF is used [58]. In this case, the weights for the EKF become the states to be estimated [26,64], and the objective is to find the optimal values for $w_{i,k}$ such that the identification error is minimized. The EKF algorithm is given by

$$\begin{aligned} M_{i,k} &= [R_{i,k} + H_{i,k}^T P_{i,k} H_{i,k}]^{-1}, \\ K_{i,k} &= P_{i,k} H_{i,k} M_{i,k}, \\ w_{i,k+1} &= w_{i,k} + \eta_i K_{i,k} e_{i,k}, \\ P_{i,k+1} &= P_{i,k} - K_{i,k} H_{i,k}^T P_{i,k} + Q_{i,k}, \end{aligned} \qquad (3.9)$$

where vector $w_{i,k}$ is the estimate of the ith weight of the ith neuron at step k. This estimate is a function of the gain K_i and the neural identification error $e_{i,k} = \chi_{i,k} - x_{i,k}$, where χ_i is the system state and x_i is the RHONN state. The gain depends on the approximate error covariance matrix P_i, a matrix of derivatives of the NN's outputs with respect to all adaptable weight parameters H_i, i.e.,

$$H_{i,k} = \left[\frac{\partial x_{i,k}}{\partial w_{i,k}} \right]^T, \tag{3.10}$$

and a scaling matrix M_i. In (3.9), Q_i is the process noise covariance matrix and R_i is the covariance matrix of the measurement noise. Additionally, a rate learning parameter η_i is introduced to modify the convergence speed, with $0 \le \eta_i \le 1$. Usually P_i, Q_i, and R_i are initialized as large diagonal matrices, with entries $P_i(0)$, $Q_i(0)$, and $R_i(0)$, respectively. Matrices Q_i and R_i are selected as fixed ones. During the neural network training, the evolutions of H_i, K_i, and P_i are guaranteed to be bounded [59].

Theorem 3.1 (Identification via RHONN [59]). *The RHONN (3.2) trained with the EKF-based algorithm (3.9) to identify the nonlinear plant (3.1) ensures that the neural identification error $e_{i,k}$ is semiglobally uniformly ultimately bounded (SGUUB); moreover, the RHONN weights remain bounded.*

For control purposes, it is of interest that the system state χ_i in (3.1) tracks a desired trajectory, which can be stated through the following inequality:

$$\| \chi_{i\delta} - \chi_i \| \le \| \chi_i - x_i \| + \| \chi_{i\delta} - x_i \|, \tag{3.11}$$

where $\|\cdot\|$ stands for the Euclidean norm and $\chi_{i\delta}$ is a bounded desired trajectory to be followed. Inequality (3.11) is established based on the separation principle for discrete-time nonlinear systems [42]; hence, from (3.11) the following requirements are stated for the neural identification and control purposes.

Requirement. We have

$$\lim_{k \to \infty} \| \chi_i - x_i \| \le \zeta_i, \tag{3.12}$$

with ζ_i being a small positive constant.

Requirement. We have

$$\lim_{k \to \infty} \| \chi_{i\delta} - x_i \| = 0. \tag{3.13}$$

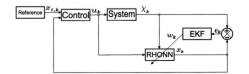

FIG. 3.2 Control scheme.

A neural identifier based on (3.5) guarantees (3.12) [59], while (3.13) is ensured by a discrete-time controller synthesized by using the sliding mode block control technique. The proposed control scheme is shown in Fig. 3.2.

Notice the stabilization of a given system can be obtained from (3.11) by considering $\chi_{i\delta} = 0$.

3.2.3 Sliding Mode Block Control Design

This subsection proposes a control law for system (3.6) through the proposed neural identifier. Considering the full state variable of the system is available from measurements, the control objective is to design a tracking control law for system (3.1). To this end, a discrete-time block control [59] and a sliding mode technique [69] are used. By using a series-parallel model (3.8), system (3.2) can be presented, possibly after a required nonlinear transformation, in the nonlinear block-controllable (NBC) form [44,52], consisting of r blocks as

$$\begin{aligned} x_{i,k+1}^1 &= W_{i,k}^1 \rho_i(\chi_{i,k}^1) + W_i'^1 \chi_{i,k}^2, \\ x_{i,k+1}^2 &= W_{i,k}^2 \rho_i(\chi_{i,k}^1, \chi_{i,k}^2) + W_i'^2 \chi_{i,k}^3, \quad (3.14) \\ &\vdots \\ x_{i,k+1}^r &= W_{i,k}^r \rho_i(\chi_{i,k}^1, \ldots, \chi_{i,k}^r) + W_i'^r u_{i,k}, \end{aligned}$$

where $x_i = \left[x_i^{1T} x_i^{2T} \cdots x_i^{rT} \right]$ is the ith block neuron state, with $x_i^j = [x_{i1}^j \ x_{i2}^j \ \ldots \ x_{il}^j]^T$, $W_{i,k}^j$ are the adjustable weight matrices, $W_i'^j$ are matrices with fixed parameters, with $j = 1, \ldots, r$ and $l = 1, \ldots, n_{ij}$, and $rank(W_i'^j) = n_{ij}$.

The block control design is started by defining the following tracking error:

$$z_{i,k}^1 = x_{i,k}^1 - \chi_{i\delta,k}^1, \tag{3.15}$$

where $\chi_{i\delta,k}^j$ is the desired trajectory signal.

Once we have defined the first new variable (3.15), the next step is taken as

$$z_{i,k+1}^1 = W_{i,k}^1 \rho_i(\chi_{i,k}^1) + W_i'^1 \chi_{i,k}^2 - \chi_{i\delta,k+1}^1. \tag{3.16}$$

System (3.16) is viewed as a block with state $z^1_{i,k}$, while the state $\chi^2_{i,k}$ is considered as a pseudocontrol input; then, desired dynamics can be imposed for this block as

$$z^1_{i,k+1} = W^1_{i,k}\,\rho_i(\chi^1_{i,k}) + W'^1_i\,\chi^2_{i,k} - \chi^1_{i\delta,k+1} = K^1_i z^1_{i,k}, \tag{3.17}$$

where $K^1_i = diag\{k^1_{i1},\ldots,k^1_{in_{i1}}\}$, with $\left|k^1_{iq}\right| < 1$, $q = 1,\ldots,n_{i1}$, to ensure the stability of (3.17), and the pseudocontrol $\chi^2_{i,k}$ is calculated as

$$\chi^2_{i\delta,k} = \left(W'^1_i\right)^{-1}\left(-W^1_{i,k}\,\rho_i(\chi^1_{i,k}) + \chi^1_{i\delta,k+1} + K^1_i z^1_{i,k}\right). \tag{3.18}$$

Note the calculated value for $\chi^2_{i\delta,k}$ in (3.18) represents the desired behavior for $\chi^2_{i,k}$, and for this reason it is denoted as $\chi^2_{i\delta,k}$.

Proceeding in a similar way as done for the first block, a second variable in the new coordinates is defined as

$$z^2_{i,k} = \chi^2_{i,k} - \chi^2_{i\delta,k}.$$

The next step in $z^2_{i,k}$ produces

$$\begin{aligned} z^2_{i,k+1} &= \chi^2_{i,k+1} - \chi^2_{i\delta,k+1} \\ &= W^2_{i,k}\,\rho_i(\chi^1_{i,k},\chi^2_{i,k}) + W'^2_i\,\chi^3_{i,k} - \chi^2_{i\delta,k+1}. \end{aligned}$$

The desired dynamics for this block is imposed as

$$\begin{aligned} z^2_{i,k+1} &= W^2_{i,k}\,\rho_i(\chi^1_{i,k},\chi^2_{i,k}) + W'^2_i\,\chi^3_{i,k} - \chi^2_{i\delta,k+1} \\ &= K^2_i z^1_{i,k}, \end{aligned} \tag{3.19}$$

where $K^2_i = diag\{k^2_{i1},\ldots,k^2_{in_{i2}}\}$ with $\left|k^2_{iq}\right| < 1$, $q = 1,\ldots,n_{i2}$.

The procedure given in the previous steps is carried out iteratively. At the last step, the known desired variable is $\chi^r_{i\delta,k}$, and the last new variable is defined as

$$z^r_{i,k} = \chi^r_{i,k} - \chi^r_{i\delta,k}.$$

As previously, by taking one step ahead, we obtain

$$z^r_{i,k+1} = W^r_{i,k}\,\rho_i(\chi^1_{i,k},\ldots,\chi^r_{i,k}) + W'^r_i\,u_{i,k} - \chi^r_{i\delta,k+1}. \tag{3.20}$$

Then, system (3.14) can be presented in the new state variables $z_i = \left[z^{1T}_i\,z^{2T}_i\cdots z^{rT}_i\right]$ as

$$\begin{aligned} z^1_{i,k+1} &= K^1_i z^1_{i,k} + W'^1_i z^2_{i,k}, \\ z^2_{i,k+1} &= K^2_i z^2_{i,k} + W'^2_i z^3_{i,k}, \\ &\ \ \vdots \\ z^{r-1}_{i,k+1} &= K^{r-1}_i z^{r-1}_{i,k} + W'^{(r-1)}_i z^r_{i,k}, \\ z^r_{i,k+1} &= W^r_{i,k}\,\rho_i(\chi^1_{i,k},\ldots,\chi^r_{i,k}) + W'^r_i u_{i,k} - \chi^r_{i\delta,k+1}. \end{aligned} \tag{3.21}$$

For a sliding mode control design [68], which takes into account a bounded control input as

$$|u_{i,k}| \le u_{i0}, \tag{3.22}$$

a sliding manifold and a control law to drive the states toward such manifold must be developed. By selecting the sliding manifold as $S_{i,k} = z^r_{i,k} = 0$, the system (3.20) can be rewritten as

$$S_{i,k+1} = W^r_{i,k}\,\rho_i(\chi^1_{i,k},\ldots,\chi^r_{i,k}) + W'^r_i u_{i,k} - \chi^r_{i\delta,k+1}. \tag{3.23}$$

Once we have defined the sliding manifold, the next step is to determine a control law, which takes into account the bound (3.22); hence, the sliding mode–based controller $u_{i,k}$ is selected as [69,68]

$$u_{i,k} = \begin{cases} u_{eq_i,k} & \text{for } \left\|u_{eq_i,k}\right\| \le u_{i0}, \\ u_{i0}\dfrac{u_{eq_i,k}}{\left\|u_{eq_i,k}\right\|} & \text{for } \left\|u_{eq_i,k}\right\| > u_{i0}, \end{cases} \tag{3.24}$$

where the equivalent control $u_{eq_i,k}$ is calculated from $S_{i,k+1} = 0$ as

$$u_{eq_i,k} = \left(W'^r_i\right)^{-1}\left(-W^r_{i,k}\,\rho_i(\chi^1_{i,k},\ldots,\chi^r_{i,k}) + \chi^r_{i\delta,k+1}\right). \tag{3.25}$$

The conditions to achieve stability of the closed-loop system (3.21) with (3.24) are given in the following theorem.

Theorem 3.2 (Sliding mode block control [52]). *Assume that*

$$u_{i0} \ge \left\|\left(W'^r_i\right)^{-1}\right\|\,\|f_{s,k}\| \tag{3.26}$$

and choose the matrix K^j_i as

$$K^j_i = diag\{k^j_{i1},\ldots,k^j_{in_{ij}}\} \text{ with } \left|k^j_{iq}\right| < 1, \tag{3.27}$$

where $f_{s,k} = -x_i^r + x_{i\delta}^r + W_{i,k}^r \rho_i(\chi_{i,k}^1, \ldots, \chi_{i,k}^r) - \chi_{i\delta,k+1}^r$ and $q = 1, \ldots, n_{ij}$. Then a solution of the closed-loop system (3.21) with (3.24) is asymptotically stable.

Reference [69] demonstrates that under condition (3.26) the state vector of the closed-loop system reaches the sliding manifold $S_{i,k} = 0$ in finite time. Then, the sliding motion on $S_{i,k} = 0$ is governed by the reduced-order *sliding mode equation* (SME)

$$
\begin{aligned}
z_{i,k+1}^1 &= K_i^1 z_{i,k}^1 + W_i'^1 z_{i,k}^2, \\
z_{i,k+1}^2 &= K_i^2 z_{i,k}^2 + W_i'^2 z_{i,k}^3, \quad (3.28)
\end{aligned}
$$

$$
\vdots
$$

$$
z_{i,k+1}^{r-1} = K_i^{r-1} z_{i,k}^{r-1}.
$$

It is easy to see that under condition (3.27), system (3.28) is asymptotically stable.

Remark 3.3. Constants k_{iq}^j in (3.27) can be arbitrarily selected, considering their magnitude values lower than 1. In this way, by arbitrarily choosing k_{iq}^j inside the interval $(-1, 1)$, the resulting system (3.28) becomes asymptotically stable. The value of such constants can be determined by the designer in accordance with the desirable system response.

Remark 3.4. It is important to note the control law (3.24)–(3.25) is bounded and ensures finite-time convergence and chattering-free motion for the closed-loop system on the manifold $S_{i,k} = z_{i,k}^r = 0$ after $r - 1$ steps. On the other hand, since the controller is based on the RNN identifier (3.14), it is a robust control technique, where the plant (3.1) model or its parameters are not required for the control law design.

Remark 3.5. Notice the identification is carried out on-line for the controller application.

3.2.4 Application for a Two–Degree of Freedom Robot

In this section, the results of Sections 3.2.2 and 3.2.3 are applied to synthesize a neural identifier–based controller for position tracking of both link 1 and link 2, in a two–degree of freedom (DOF) planar rigid robot (Fig. 3.3).

3.2.4.1 Robot Model Description
The robot dynamics [19] is given by

$$
\ddot{\Theta} = M^{-1}(\Theta)\left[\tau - V(\Theta, \dot{\Theta}) - F(\Theta, \dot{\Theta})\right], \quad (3.29)
$$

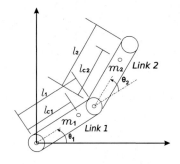

FIG. 3.3 Two-DOF planar robot.

where

$$
M(\Theta) = \begin{bmatrix} D_{11}(\Theta) & D_{12}(\Theta) \\ D_{12}(\Theta) & D_{22}(\Theta) \end{bmatrix}, \quad (3.30)
$$

$$
V(\Theta, \dot{\Theta}) = \begin{bmatrix} V_1(\Theta, \dot{\Theta}) \\ V_2(\Theta, \dot{\Theta}) \end{bmatrix}, \quad (3.31)
$$

$$
F(\Theta, \dot{\Theta}) = \begin{bmatrix} F_1(\Theta, \dot{\Theta}) \\ F_2(\Theta, \dot{\Theta}) \end{bmatrix}, \quad (3.32)
$$

with entries of (3.30)–(3.32) as
$$
D_{11}(\Theta) = m_1 l_{c1}^2 + m_2(l_1^2 + l_{c2}^2 + l_1 l_{c2} c_2) + I_{zz_1} + I_{zz_2},
$$
$$
D_{12}(\Theta) = m_2 l_{c2}^2 + m_2 l_1 l_{c2} c_2 + I_{zz_2},
$$
$$
D_{22}(\Theta) = m_2 l_{c2}^2 + I_{zz_2},
$$
$$
V_1(\Theta, \dot{\Theta}) = -m_2 l_1 l_{c2} s_2 (\dot{\theta}_1 + \dot{\theta}_2)\dot{\theta}_2 - m_2 l_1 l_{c2} \dot{\theta}_1 \dot{\theta}_2 s_2,
$$
$$
V_2(\Theta, \dot{\Theta}) = m_2 l_1 l_{c2} s_2 (\dot{\theta}_1)^2,
$$
$$
F_1(\Theta, \dot{\Theta}) = \mu_1 \dot{\theta}_1,
$$
$$
F_2(\Theta, \dot{\Theta}) = \mu_2 \dot{\theta}_2,
$$
where $\Theta = [\theta_1 \ \theta_2]^T$ is the links' position vector, $\dot{\Theta} = [\dot{\theta}_1 \ \dot{\theta}_2]^T$ is the velocity vector, $\tau = [\tau_1 \ \tau_2]^T$ is the applied torque vector, $s_2 = sen(\theta_2)$, and $c_2 = cos(\theta_2)$.

By defining the state variables as

$$
\chi_1^1 = \theta_1, \quad \chi_2^1 = \theta_2, \quad \chi_1^2 = \dot{\theta}_1, \quad \chi_2^2 = \dot{\theta}_2,
$$

an Euler discretization is carried out for system (3.29), resulting in

$$
\begin{aligned}
\chi_{1,k+1}^1 &= \chi_{1,k}^1 + \chi_{1,k}^2 T, \\
\chi_{2,k+1}^1 &= \chi_{2,k}^1 + \chi_{2,k}^2 T, \\
\chi_{1,k+1}^2 &= \chi_{1,k}^2 + \left(\frac{-D_{22}(V_1 + F_1) + D_{12}(V_2 + F_2)}{D_{11}D_{22} - D_{12}^2} \right. \\
&\quad \left. + \frac{D_{22}u_1 - D_{12}u_2}{D_{11}D_{22} - D_{12}^2} \right) T, \quad (3.33) \\
\chi_{2,k+1}^2 &= \chi_{2,k}^2 + \left(\frac{D_{12}(V_1 + F_1) - D_{11}(V_2 + F_2)}{D_{11}D_{22} - D_{12}^2} \right. \\
&\quad \left. + \frac{-D_{12}u_1 + D_{11}u_2}{D_{11}D_{22} - D_{12}^2} \right) T,
\end{aligned}
$$

where T is the sampling time and $u_1 = \tau_1$ and $u_2 = \tau_2$ are the applied torques. Note that system (3.33) is already in the NBC form.

Remark 3.6. System (3.33) is used only for simulation purposes, and in addition, its structure is employed to design the neural identifier; nonetheless, the system parameters are always assumed to be unknown in the design.

3.2.4.2 Neural Network in the Controllable Form

To identify the assumed uncertain robot model, from (3.14) the following series-parallel neural network is proposed:

$$
\begin{aligned}
x^1_{i,k+1} &= w^1_{i1,k} S_i(\chi^1_{i,k}) + w^1_{i2,k} + w'^1_i \chi^2_{i,k}, \\
x^2_{i,k+1} &= w^2_{i1,k} S_i(\chi^2_{i,k}) + w^2_{i2,k} + w'^2_i u_i, \quad (3.34)
\end{aligned}
$$

where $x^1_{i,k}$ identifies to $\chi^1_{i,k}$ and in a similar way $x^2_{i,k}$ identifies to $\chi^2_{i,k}$, with $i = 1, 2$; w^j_{ip} are the adaptable weights, p is the number of adjustable weights with $p = 1, 2$, $j = 1, 2$, and w'^j_i are fixed weights. To adapt the neural weights, the training algorithm (3.9) is employed.

3.2.4.3 Controller Design

The control objective is that the angular position $x^1_{i,k}$ tracks a desired reference signal $\chi^j_{i\delta,k}$, which is achieved by synthesizing a controller based on the sliding mode technique, as described in Section 3.2.3. Firstly, the tracking error is defined as

$$
z^1_{i,k} = x^1_{i,k} - \chi^1_{i\delta,k}.
$$

Secondly, using (3.34) and introducing the desired dynamics for $z^1_{i,k}$, the tracking error results in

$$
\begin{aligned}
z^1_{i,k+1} &= w^1_{i1,k} S_i(\chi^1_{i,k}) + w^1_{i2,k} + w'^1_i \chi^2_{i,k} - \chi^1_{i\delta,k+1} \\
&= k^1_i z^1_{i,k} \quad (3.35)
\end{aligned}
$$

with $\left| k^1_i \right| < 1$.

The desired value $\chi^2_{i\delta,k}$ for the pseudocontrol input $\chi^2_{i,k}$ is calculated from (3.35) as

$$
\chi^2_{i\delta,k} = \left(w'^1_i \right)^{-1}
$$
$$
\times \left(-w^1_{i1,k} S_i(\chi^1_{i,k}) - w^1_{i2,k} + \chi^1_{i\delta,k+1} + k^1_i z^1_{i,k} \right).
$$

At the second step, a new variable is introduced as

$$
z^2_{i,k} = x^2_{i,k} - \chi^2_{i\delta,k}.
$$

In the next step, one obtains

$$
z^2_{i,k+1} = w^2_{i1,k} S_i(\chi^2_{i,k}) + w^2_{i2,k} + w'^2_i u_i - \chi^2_{i\delta,k+1}. \quad (3.36)
$$

Now, system (3.34) with the new variables $z^1_{i,k}$ and $z^2_{i,k}$ can be presented as

$$
\begin{aligned}
z^1_{i,k+1} &= k^1_i z^1_{i,k} + w'^1_i z^2_{i,k}, \\
z^2_{i,k+1} &= w^2_{i1,k} S_i(\chi^2_{i,k}) + w^2_{i2,k} + w'^2_i u_i - \chi^2_{i\delta,k+1}. \quad (3.37)
\end{aligned}
$$

The sliding manifold is selected as $S_{i,k} = z^2_{i,k} = 0$, while the control law is given in (3.24), where

$$
\begin{aligned}
u_{eq_i,k} = &\left(w'^2_i \right)^{-1} \\
&\times \left(-w^2_{i1,k} S_i(\chi^2_{i,k}) - w^2_{i2,k} + \chi^2_{i\delta,k+1} \right),
\end{aligned}
$$

which is obtained from calculating $S_{i,k+1} = 0$.

A motion on $S_{i,k} = 0$ is described by the following SME:

$$
z^1_{i,k+1} = k^1_i z^1_{i,k},
$$

which is asymptotically stable for

$$
\left| k^1_i \right| < 1.
$$

3.2.4.4 Simulation Results

The plant parameters used for the simulation are given in Table 3.1.

The reference signals to be tracked are given by

$$
\begin{aligned}
\chi^1_{1\delta,k} &= 1.4 \sin(0.25 \, k \, T) \, rad, \\
\chi^1_{2\delta,k} &= 1.4 \sin(0.34 \, k \, T) \, rad.
\end{aligned}
$$

The simulations are performed using *Matlab/ Simulink*, a trademark of *MathWorks, Inc.* The trajectory tracking performance for link 1 ($i = 1$) and link 2 ($i = 2$) positions are shown in Fig. 3.4.

3.2.4.5 Experimental Results

The neural identification and control schemes are experimentally evaluated. The real-time initial conditions are given in Table 3.2; the identifier (3.34) and controller

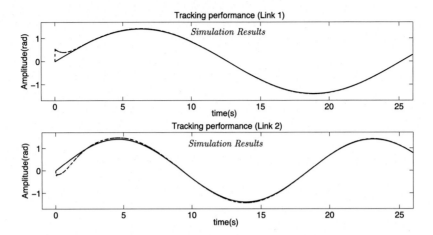

FIG. 3.4 Tracking performance simulation of $\chi_{i\delta}^1$ (*solid line*), χ_i^1 (*dash-dot*), and x_i^1 (*dashed line*).

TABLE 3.1
Plant parameters.

Parameter	Value	Description
l_1	0.3 m	Length of the link 1
l_{c1}	0.2 m	Mean length of the link 1
l_2	0.25 m	Length of the link 2
l_{c2}	0.1 m	Mean length of the link 2
m_1	1 kg	Mass of the link 1
m_2	0.3 kg	Mass of the link 2
I_{zz1}	0.05 kg·m²	Moment of inertia 1
I_{zz2}	0.004 kg·m²	Moment of inertia 2
μ_1	0.005 kg/s	Friction coefficient 1
μ_2	0.0047 kg/s	Friction coefficient 2

TABLE 3.2
Real-time initial conditions.

$\chi_{1,0}^1$	0.25 rad	$\chi_{2,0}^1$	0.3 rad
$\chi_{1,0}^2$	0 rad/s	$\chi_{2,0}^2$	0 rad/s
$x_{i,0}^1$	0 rad	$x_{i,0}^2$	0 rad/s

TABLE 3.3
Identifier and controller parameters.

Parameter	Value	Parameter	Value
$w_1'^1$	0.008	$w_2'^1$	0.008
$w_1'^2$	0.7	$w_2'^2$	0.5
k_1^1	0.73	k_2^1	0.75
η_i^1	0.8	η_i^2	0.85
u_{10}	40 Nm	u_{20}	48 Nm

to the one given for the RHONN. The weights w_{ip}^j are selected in a random way. In the experimental results, the initial conditions for the RHONN are given as zero.

3.3 NEURAL MODELING AND CONTROL FOR CONTINUOUS-TIME SYSTEMS

In a similar way as presented in Section 3.2, the neural methodology to identify and control continuous-time uncertain nonlinear systems is described in this section. In this case, a polynomial NN is used.

3.3.1 Continuous-Time Uncertain Nonlinear Systems

In this section the problem of approximating a disturbed and uncertain nonlinear system with dynamical behavior given by

$$\dot{\mathcal{X}} = \mathcal{F}(\mathcal{X}, u, t) + \bar{\Gamma}, \qquad (3.38)$$
$$\mathcal{y} = \mathcal{C}(\mathcal{X}) \qquad (3.39)$$

parameters are shown in Table 3.3. The sample time is fixed at $T = 0.008$ s.

The experimental tracking performances of the position for link 1 and link 2 are shown in Fig. 3.5 and Fig. 3.6, respectively (in the bottom pictures, a zoom of the transient of the tracking performance is shown). Note the plant initial conditions are selected different

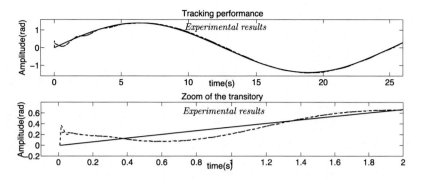

FIG. 3.5 Experimental tracking performance of $\chi^1_{1\delta}$ (*solid line*), χ^1_1 (*dash-dot*) and x^1_1 (*dashed line*).

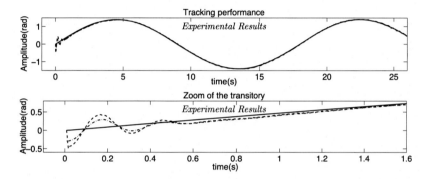

FIG. 3.6 Experimental tracking performance of $\chi^1_{2\delta}$ (*solid line*), χ^1_2 (*dash-dot*) and x^1_2 (*dashed line*).

is going to be addressed. In (3.39), $\mathcal{X} \in \mathbb{R}^n$ is the actual system state, $u \in \mathbb{R}^m$ is the system input, and $\mathcal{Y} \in \mathbb{R}^p$ is the system output; $\mathcal{F}$ and $\mathcal{C}$ are smooth vector fields of appropriate dimensions that are unknown or partially known, and $\bar{\Gamma}$ is a term representing unknown and bounded disturbances, uncertainties, and/or dynamics that are not modeled. Here, an adaptive polynomial neural identifier is proposed to obtain the dynamical behavior of the system (3.38)–(3.39). In order to do this, the system is assumed to be observable; in consequence, the identification process can be carried out. Additionally, let us consider that system (3.38)–(3.39) is controllable and observable [31].

3.3.2 Polynomial Neural Identifier

The system given by (3.38)–(3.39) may be approximated by an NN-based identifier written as

$$\dot{x} = f(x)\theta + Bu + \Gamma(\theta), \qquad (3.40)$$
$$y = Cx, \qquad (3.41)$$

where $x \in \mathbb{R}^n$ is the identifier state vector, $u \in \mathbb{R}^m$ is the control input, and $y \in \mathbb{R}^p$ is the identifier output; func-

tion $f(x)$ is a smooth map of appropriate dimension. It is proposed for the map to be polynomial with respect to x. In addition, B and C are the input and output matrices, respectively, of appropriate dimensions; θ is the identifier weight vector and it is adapted online until the identifier converges toward the system (3.38)–(3.39). Finally, the term $\Gamma(\theta)$ may represent additional (constant or slowly time-varying) parameters of the identifier. The main motivation for considering the $f(x)$ polynomial in x is that polynomials are amenable for algebraic manipulation and they may be orthogonal, which would imply that better approximations are achieved while increasing the number of factors. Accordingly, different polynomial bases for $f(x)$ can be used to approximate the vector field $\mathcal{F}$ in (3.38), such as Chebyshev polynomials [45] and Legendre polynomials [23]. Additionally, there are different nonlinear systems with a natural polynomial structure [16,10,13,24,54,27,70].

Remark 3.7. Another reason for considering a polynomial structure, specifically for function $f(x)\theta$ in (3.40), is that it always admits the state-dependent coefficient factorization[1] as $f(x)\theta = A(x,\theta)x$. According to this, system (3.40)–(3.41) can be rewritten as

$$\dot{x} = A(x,\theta)x + Bu + \Gamma(\theta), \quad (3.42)$$

$$y = Cx, \quad (3.43)$$

where the controllability and observability properties must be proven, as described in [8,53,27].

Remark 3.8. Under the assumption the real system is controllable, matrix B in (3.42) is proposed such that the identifier is controllable. Note that the matrix B is designed from the knowledge of the real system and obtained using a gray-box methodology.

It is worth noting that $f(x)\theta$ is linear with respect to the entries of vector θ. The main reason for doing this is that different learning algorithms that exploit such linear dependence can be used.

3.3.2.1 PNN Learning Algorithm

Many algorithms can be used for adjusting neural weights; however, least squares (LS) are perhaps the most successful ones regardless the notion that neural structures can be derived from different methods. In this chapter, for the online adaptation of the identifier weights θ, an LS algorithm is applied, the so-called *Kalman–Bucy filter* (KBF) [62], used as a state estimator [11], where the weights become the states to be estimated. The main objective of the KBF is to determine the optimal values for the weight vector θ such that the identification error

$$\varepsilon = x - \mathcal{X} \quad (3.44)$$

is minimized.

Here, it is assumed that an ideal unknown weights vector θ^* produces the minimum identification error, whose dynamics can be described by

$$\dot{\theta}^* = 0 \quad (3.45)$$

with output

$$y^* = h(x^*,\theta^*), \quad (3.46)$$

where the ideal state vector x^* depends on θ^* in accordance with its respective dynamics from (3.40). Assuming that (3.45) and (3.46) are affected by uncorrelated zero mean white Gaussian noises of spectral intensities

---

[1]For instance, the polynomial scalar system $\dot{x} = -x + x^3$ can be factorized as $\dot{x} = a(x)x$, with $a(x) = (x^2 - 1)$.

$\Psi \in \mathbb{R}^{2n \times 2n}$ and $1/g \in \mathbb{R}$, respectively, the KBF is given as [62]

$$\dot{\theta} = -g\,\Phi\,w_b\,\varepsilon,$$

$$\dot{\Phi} = \Psi - g\,\Phi\,w_b\,w_b^T\,\Phi, \quad \Psi,\ g > 0, \quad (3.47)$$

where θ is the estimated value for θ^* and vector w_b contains the selected polynomial basis to approximate the vector fields in (3.38)–(3.39). Note that the constant values of Ψ and g become design parameters used to ensure the identification error convergence. Matrix Φ is of appropriate dimension, is named the covariance matrix, and acts in the θ updating law as a directional adaption gain. The initial condition for Φ is $\Phi(0) > 0$, whereas for θ it is arbitrary; $\Phi(0)$ is usually chosen to reflect the confidence in the initial estimate of $\theta(0)$. It is recommended to select a large value for $\Phi(0)$ to ensure better performance of the identification process. Note that the adaptation law (3.47) is applied for each system state variable.

3.3.3 Nonlinear Optimal Neural Control Design

On the basis of the state-dependent representation for (3.42)–(3.43), it is possible to synthesize an optimal controller and to obtain an analytical solution via the Riccati equation. In this chapter, for the optimal controller design, the output of the system is required to track a desired trajectory as close as possible in an optimal sense and with minimum control effort expenditure [3,5,38,40]. Hence, the trajectory tracking error is defined as

$$e = r - y$$
$$= r - Cx, \quad (3.48)$$

where r is the desired reference to be tracked by the system output y.

The quadratic cost functional J to be minimized, associated with system (3.42)–(3.43), is defined as

$$J = \frac{1}{2}\int_{t_0}^{\infty}\left(e^T Q e + u^T R u\right)dt, \quad (3.49)$$

where Q and R are symmetric and positive definite matrices. Therefore, the optimal tracking solution occurs when the control u is such that the criterion (3.49) is minimized.

3.3.3.1 Optimal Tracking Controller

Firstly, an approach to establish an optimal tracking control solution is done assuming the nonexistence of the disturbance (i.e., $\Gamma(\theta) = 0$) for (3.42) and omitting the parameter dependence of θ to simplify the notation in all system functions. A posteriori, the robust optimal tracking control, i.e., one that considers system disturbance, is stated in the next subsection.

By considering that the state is available for feedback, the optimal tracking solution is established as follows.

Theorem 3.3 (Nonlinear optimal tracking controller [53]). *Assume that system (3.42), with $\Gamma(\theta) = 0$, is state-dependent controllable and state-dependent observable. Then the optimal control law*

$$u^*(x) = -R^{-1}B^T(x)\,(P(x)\,x - z(x)) \qquad (3.50)$$

achieves trajectory tracking for system (3.42) along the desired trajectory r, where $P(x)$ is the solution of the equation

$$\begin{aligned}\dot{P}(x) \;=\;& -C^T(x)\,Q\,C(x) + P(x)\,B(x)\,R^{-1}\,B^T(x)\\ &\times P(x) - A^T(x)\,P(x) - P(x)\,A(x)\end{aligned} \qquad (3.51)$$

and $z(x)$ is the solution of the equation

$$\begin{aligned}\dot{z}(x) \;=\;& -\Big[A(x) - B(x)\,R^{-1}B^T(x)\,P(x)\Big]^T z(x)\\ &-C^T(x)\,Q\,r\end{aligned} \qquad (3.52)$$

with boundary conditions $P(x(\infty)) = 0$ and $z(x(\infty)) = 0$, respectively. Control law (3.50) is optimal in the sense that it minimizes the cost functional (3.49), which has an optimal value function given as

$$J^* = \frac{1}{2}x^T(t_0)\,P(x(t_0))\,x(t_0) - z^T(x(t_0))\,x(t_0) + \varphi(t_0), \qquad (3.53)$$

where φ is the solution to the scalar differentiable function

$$\dot{\varphi} = -\frac{1}{2}r^T\,Q\,r + \frac{1}{2}z^T\,B(x)\,R^{-1}B^T(x)\,z \qquad (3.54)$$

with $\varphi(\infty) = 0$.

Matrix (3.51) and vector (3.52) need to be solved backward in time; nonetheless, in [40] and [73] a change of variable is presented such that these equations can be solved forward in time. This is done by multiplying a minus sign with the right-hand side of (3.51) and (3.52).

3.3.3.2 Robust Optimal Tracking Controller

To improve performance, the controller design must consider the case when the disturbance term $\Gamma(\theta)$ is actually part of system (3.42). In this case, an integral term

of the tracking error e is included such that this disturbance is rejected. Accordingly, the integral term for the tracking error is defined as

$$\dot{q} = -e, \qquad (3.55)$$

where $q \in \mathbb{R}^p$ is a state vector of integrators for a system with p outputs. Then, an augmented system which includes the integrator can be established as

$$\begin{aligned}\dot{x}_a \;=\;& \begin{bmatrix} \dot{q} \\ \dot{x} \end{bmatrix}\\[4pt] =\;& \begin{bmatrix} -e \\ A(x)x + B(x)u + \Gamma \end{bmatrix}\\[4pt] =\;& \begin{bmatrix} C(x)x - r \\ A(x)x + B(x)u + \Gamma \end{bmatrix},\end{aligned} \qquad (3.56)$$

with $x_a = [q^T,\ x^T]^T$. The dependence of weight θ in all functions is omitted for simplicity of notation. System (3.56) can be rewritten as

$$\begin{aligned}\dot{x}_a &= A_a(x_a)\,x_a + B_a(x_a)\,u + D_a, &(3.57)\\ y &= C_a x_a, &(3.58)\end{aligned}$$

where $A_a(x_a) = \begin{bmatrix} 0 & C(x) \\ 0 & A(x) \end{bmatrix}$, $B_a(x_a) = \begin{bmatrix} 0 \\ B(x) \end{bmatrix}$, $C_a = \begin{bmatrix} 0 & C \end{bmatrix}$, and $D_a = \begin{bmatrix} -r \\ \Gamma \end{bmatrix}$. For system (3.57), let us consider the problem of minimizing the cost functional

$$J = \frac{1}{2}\int_{t_0}^{\infty} \Big(q^T\,Q_I\,q + e^T\,Q\,e + u^T\,R\,u\Big)\,dt, \qquad (3.59)$$

where Q_I is a parameter weighting the integrator performance, which can be considered as the integrator gain, Q is a matrix weighting the time evolution of the error, and R is a matrix weighting the control effort expenditure. These matrices are used to establish a trade-off between state performance and control effort. The robust optimal tracking solution is established as follows:

$$u^*(x_a) = -R^{-1}B_a^T(x_a)\,(P_a(x_a)\,x_a - z_a(x_a)), \qquad (3.60)$$

with $P_a(x_a)$ and $z_a(x_a)$ defined in accordance with Section 3.3.3.1.

Remark 3.9. It is worth mentioning that the objective of the identifier is to capture most of the dynamical behavior of the unknown nonlinear system by means of an adaptive polynomial model, but it is not based on physical laws, and therefore it does not intend to determine the real system parameters or the real model.

3.3.4 Application to a Glucose–Insulin System

In this section the above ideas are applied to control a system with an assumed unknown mathematical model.

3.3.4.1 Bergman Model

The Bergman minimal model (BeMM) is a system that describes the dynamics of the glucose uptake after an external stimulus in humans. It consists of two parts: one that describes the changes of the glucose and another that describes the dynamics of the pancreatic insulin release in response to the glucose stimulus. Here, this model is only used to simulate the glucose–insulin dynamics, which is described by [12]

$$\dot{G} = -p_1 G - XG + p_1 G_b + D, \quad (3.61)$$

$$\dot{X} = -p_2 X + p_3 (I - I_b), \quad (3.62)$$

$$\dot{I} = -\eta(I - I_b) + \gamma(G - h)t + u, \quad (3.63)$$

where G, X, and I are the plasma glucose concentration, the insulin influence on glucose concentration reduction, and the insulin concentration in plasma, respectively, thus $\mathcal{X} = [G \ X \ I]^T$. The control input u represents the insulin infusion rate, p_1 is the insulin-independent glucose utilization rate, p_2 is the rate of decrease of the tissue glucose uptake ability, p_3 is the insulin-dependent increase of the glucose uptake ability, and $D = \dfrac{D_G A_G t e^{-t/T_{maxI}}}{V_G T_{maxG}^2}$ is the disturbance caused by the meal [29], where D_G is the meal carbohydrate load, A_G is the carbohydrate bioavailability, T_{maxI} is the time-to-maximum insulin absorption, T_{maxG} is the time-of-maximum appearance rate of glucose in the accessible glucose compartment, and V_G is the glucose distribution space. The term $\gamma(G - h)t$ represents the pancreatic insulin secretion after a meal intake at $t = 0$. The threshold value of glucose above which the pancreatic β-cells release insulin is represented by h and γ is the rate of the pancreatic β-cells' release of insulin after the glucose injection. Because this section is focused on insulin therapy, which is usually given to type 1 diabetic patients, γ is assumed to be zero, to represent the true dynamic of this disease [21]. The parameter η is the first-order decay rate for insulin in blood.

3.3.4.2 Polynomial Neural Identification for the Bergman Minimal Model

Note that system (3.61)–(3.63) has a polynomial structure and it becomes natural to fit an identifier that exploits this structure. Then, the identifier is proposed as

$$\dot{x}_1 = \theta_1 x_1 - x_2 x_1 + \theta_2, \quad (3.64)$$

$$\dot{x}_2 = \theta_3 x_2 + \theta_4 x_3 + \theta_5, \quad (3.65)$$

$$\dot{x}_3 = \theta_6 x_3 + \theta_7 + u, \quad (3.66)$$

where the state vector $x = [x_1 \ x_2 \ x_3]^T$ identifies, state-to-state, to the assumed actual state vector $\mathcal{X} = [G \ X \ I]^T$.

System (3.64)–(3.66) can be presented as (3.42)–(3.43), with $A(x, \theta) = \begin{bmatrix} \theta_1 & -x_1 & 0 \\ 0 & \theta_3 & \theta_4 \\ 0 & 0 & \theta_6 \end{bmatrix}$, $B = \begin{bmatrix} 0 \\ 0 \\ 1 \end{bmatrix}$,

$C = \begin{bmatrix} 1 & 0 & 0 \end{bmatrix}$, and $\Gamma(\theta) = \begin{bmatrix} \theta_2 \\ \theta_5 \\ \theta_7 \end{bmatrix}$, where $\theta = [\theta_1 \ \theta_2 \ \theta_3 \ \theta_4 \ \theta_5 \ \theta_6 \ \theta_7]^T$ are the weights to be identified by the Kalman–Bucy algorithm by using the identification error $\varepsilon = x_1 - G$. Therefore, the robust optimal tracking control can be applied.

3.3.4.3 Robust Optimal Tracking Control Applied to the Bergman Model

The output of the Bergman model (BeM) is the blood glucose level (G). It is required to add only one integrator; hence, the augmented system becomes

$$\dot{x}_a = \begin{bmatrix} -e \\ A(x, \theta)x + Bu + \Gamma \end{bmatrix}$$

$$= \begin{bmatrix} x_1 - r \\ A(x, \theta)x + Bu + \Gamma \end{bmatrix}, \quad (3.67)$$

with $x_a = [q \ x]^T = [q \ x_1 \ x_2 \ x_3]^T$, which can be rewritten as

$$\dot{x}_a = A_a(x_a, \theta)x_a + B_a u + \Gamma_a, \quad (3.68)$$

$$y = C_a x_a, \quad (3.69)$$

where $A_a(x_a, \theta) = \begin{bmatrix} 0 & 1 & 0 & 0 \\ 0 & \theta_1 & -x_1 & 0 \\ 0 & 0 & \theta_3 & \theta_4 \\ 0 & 0 & 0 & \theta_6 \end{bmatrix}$, $B_a = \begin{bmatrix} 0 \\ 0 \\ 0 \\ 1 \end{bmatrix}$,

$C_a = \begin{bmatrix} 0 & 1 & 0 & 0 \end{bmatrix}$, and $\Gamma_a = \begin{bmatrix} -r \\ \theta_2 \\ \theta_5 \\ \theta_7 \end{bmatrix}$, with r being the reference value for the glucose level, and the cost functional to be minimized is (3.59). For the augmented system, the optimal controller is given by (3.60).

3.3.4.4 Simulation Results

Both the identifier and the controller designed above are used in simulation experiments, where their effectiveness can be better appreciated. The parameters used

TABLE 3.4
Bergman model parameters.

Parameter	Value	Unit
p_1	0.1082	1/min
p_2	0.02	1/min
p_3	5.3×10^{-6}	ml/μUmin^2
η	0.2659	1/min
G_b	110	mg/dl
I_b	90	μU/ml
D_G	125	mg
V_G	13.79	dl
A_G	0.8	nondimensional
T_{maxG}	5	min
T_{maxI}	15	min

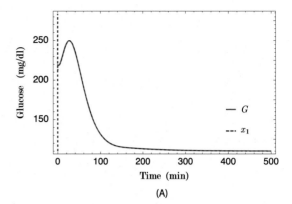

(A)

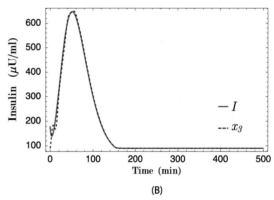

(B)

FIG. 3.7 Identification process for the Bergman model. **(A)** Optimal identification for glucose. **(B)** Optimal identification for insulin.

in the simulation for the (BeM) are given in Table 3.4; they are taken from [35]. The initial conditions for the BeM are $G(0) = 1.2G_b$, $I(0) = 1.2I_b$, and $X(0) = 0$. In the process of identification the basis w_b in (3.47) are $w_1 = [x_2\ 1]^T$, $w_2 = [x_3\ x_4\ 1]^T$ and $w_3 = [x_4\ 1]^T$, which allows the identification process to be more efficient. The identifier parameters to ensure the identification convergence are $\Psi_1 = diag\{0.05,\ 80\}$, $\Psi_2 = diag\{0.01,\ 0.01,\ 0.01\}$, $\Psi_3 = diag\{1,\ 100000\}$, $g_1 = 100$, $g_2 = 90000000$, and $g_3 = 5050000$. Once the identifier has converged, the optimal control law is applied, which happens at time $t \geq 80$ min. The parameters for the robust optimal tracking controller, which determine the convergence ratio of the error through the control law, are $Q_I = 0.09$, $Q = 1.5$, and $R = 1$. These parameter values are selected such that an adequate performance of the control system is achieved.

Fig. 3.7A shows the identification for the basal glucose response, corresponding to the BeM variable (G), by means of the proposed identifier variable x_1. Fig. 3.7B shows the identification of the basal insulin response, corresponding to the BeM variable (I), by means of x_3.

Fig. 3.8A shows the glucose level regulation for a reference level of $r = 110$ mg/dl. Fig. 3.8B shows the control signal (u) that represents the required level of exogenous insulin to keep the blood glucose on the reference level established for a type 1 diabetic patient. Finally, the capabilities of the optimal control scheme for regulating the glucose level to different reference values at different time intervals (r) are illustrated in Fig. 3.9, i.e., for $t < 400$ min the reference level is $r = 115$ mg/dl, for $400 \leq t < 600$ min the reference level is

$r = 105$ mg/dl, for $600 \leq t < 800$ min the reference level is $r = 115$ mg/dl, and in the last time interval, $t \geq 800$ min, the reference level is $r = 110$ mg/dl.

3.4 FURTHER NN APPLICATIONS

NNs have found different and important applications in engineering and control systems. In the following, a few of those applications are described.

3.4.1 Reduced-Order Models

Over the last years, investigators have attempted to develop several methodologies to reduce the order in nonlinear systems [20]. These methods work exceedingly well when the dynamics of the system can be considered to be linear. Researchers have in recent years looked into the creation of black-box models, which use only input and output data to create a model which can be used to predict system behavior. Many such methods, a majority of which are based on the seminal work done by Kalman [34], are used in the structural dy-

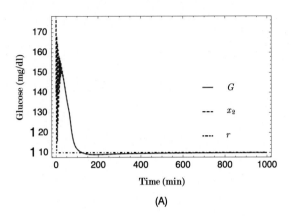

(A)

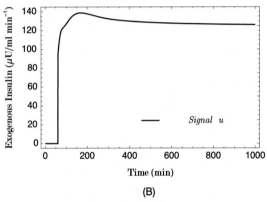

(B)

FIG. 3.8 Identification and optimal tracking control for the Bergman model. **(A)** Glucose signal at reference level r. **(B)** Control signal to keep the glucose at reference level r.

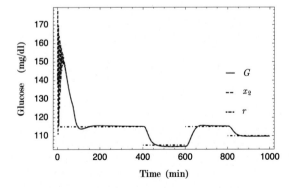

FIG. 3.9 Identification and optimal tracking control for the Bergman model to maintain the glucose level at different reference levels.

namics community to build linear state-space equations and to predict system parameters [32,33]. Recently similar methods have been used to build linear state-space aerodynamics models as well [67,18]. In [6], a nonlinear system identification methodology is presented which is used to identify a set of low-dimensional, nonlinear, first-order ordinary differential equations.

The description of a dynamic system can be obtained from mathematical models (described by differential or difference equations) or experimental results. Based on a priori knowledge about the system, the differences between the high-order model dynamics and the reduced-order model can be taken into account as disturbances or dynamics rapidly disappear, which can be handled using adaptive identification methodologies. In any case, the model reduction procedures might be flexible enough to let the user indicate the essential behaviors that need to be captured for the corresponding application [6]. In the order reduction process, the dynamics of a system model M of high order n is approximated by a model M_r of order $r < n$. Assume that (3.38) is a supposedly uncertain and disturbed nonlinear system with a complete order, that is, $\mathcal{X} \in \mathbb{R}^n$. Then a reduced-order model can be proposed as

$$\dot{x} = f(x)\theta + Bu + \Gamma(\theta), \qquad (3.70)$$
$$y = Cx, \qquad (3.71)$$

where $x \in \mathbb{R}^r$, with $r < n$, and θ is the identifier weights' vector.

The quality of the approximation is usually evaluated by looking at the model reduction error, that is, the difference between the outputs of the original system and the outputs of the reduced-order model, driven by the same excitation input. Model order reduction is a branch of systems and control theory which studies properties of dynamical systems in applications for reducing their complexity, while preserving their input–output behavior. To ensure an adequate approximation of the original system, it is needed to preserve basic system properties, essential dynamics, stability, etc. Proposing reduced-order models to approximate the essential behaviors of uncertain and disturbed nonlinear systems has several advantages, since fewer parameters need to be adapted than with complete-order identifiers. The reduced-order modeling allows a faster convergence and lower dimension, and some variables can be considered as disturbances.

3.4.2 Observers Design

For implementing control systems and in order to obtain a high performance from a synthesized controller, the knowledge of the system model is usually required, where the state variables involved in the system are used for feedback. Hence, under this condition, it is assumed that the full state variables are available for feedback; however, in practice, when the full state vector is not accessible from measurements, synthesizing a state observer in order to estimate the missing state variables is necessary. This estimation can be derived from the synthesis of a full-order observer (when the order of the observer is the same as that of the system) or a reduced-order observer (when the order of the observer is lower than that of the system). State estimation has been studied by many authors, who have obtained interesting results in different directions ([1,7,9,22,65,66] and references therein), where uncertainties are not always considered [41,43,72]. Usually, an observer design is based on the system model, by taking into account both the input–output relationships and the system dynamics, such that the observer estimates the unmeasured variables; nonetheless, when the system model is uncertain, disturbed, or even unknown, the design of robust observers is necessary, ensuring the state estimation of such system [15,17,30,71]. It is worth remarking there always exist external and internal uncertainties in real systems.

In this sense, different neural schemes have emerged to develop robust observers for systems which are uncertain or unknown [2,37,39,46,55,56]. Additionally, the resulting NN-based observers can be used as mathematical models for unknown linear or nonlinear systems. Hence, NNs have become a well-established methodology for solving difficult problems in engineering related to modeling (neural identification) and controller and observer designs in general nonlinear and complex systems.

3.4.3 Prediction

One of the marvels of the technological development is the increasing capability of measuring and storing data about all kinds of industrial, scientific, economic, and societal processes. These capabilities have unleashed important artificial intelligence areas such as big data and time series forecasting. Indeed, with the availability of time series records, much decision taking has been substantially improved. Think, for instance, about the volumes of water that can be used for electricity generation without jeopardizing agricultural uses of the liquid. Also, prices of commodities or assets produce what is named time series. Different kinds of financial time series have been recorded and studied for decades.

Nowadays, all transactions on the financial market are recorded, leading to a huge amount of data available, either for free on the internet or commercially. Financial time series analysis is of great interest to practitioners as well as to theoreticians, for making inferences and predictions. The goal of time series forecasting can be stated succinctly as follows: given a sequence $y(1), y(2), ..., y(N)$ up to time N, find the continuation $y(N + 1), y(N + 2), ..., y(N + M)$ up to time M. The series may arise from sampling a continuous-time system and be either stochastic or deterministic.

Due to their nonlinear modeling characteristics, artificial NNs have been successfully applied not only in control design but in many technical areas such as pattern classification, speech recognition, deep learning, and certainly time series prediction. For instance, in [60] higher-order NNs have been applied to the intricate problem of electricity demand and price. Also, artificial NNs have the well-known advantages of being able to approximate nonlinear functions and being able to solve problems where the input–output relationship is neither well defined nor easily computable, because NNs are data-driven. Accordingly, in a world that has increased the storing and measuring capabilities, NNs are expected to play a predominant role in the applications of artificial intelligence.

3.5 CONCLUSIONS

This chapter has presented a methodology for modeling and controlling uncertain nonlinear systems, as in the discrete-time as in the continuous-time framework, through the use of NNs, which are trained online using Kalman filters. By employing this methodology, explicit knowledge of a real system is not necessary. The neural approach can be used when no information of a system is available (only input–output data relationships) or when a system is difficult to model using traditional methods (using physical laws). The properties and flexibility of the neural designs are mathematically stated and evidenced through examples. Simulation results show the control goals are achieved, i.e., the proposed controllers ensure the stability and trajectory tracking for the variables in uncertain systems.

REFERENCES

1. A.Y. Alanis, E. Rangel, J. Rivera, N. Arana-Daniel, C. Lopez-Franco, Particle swarm based approach of a real-time discrete neural identifier for linear induction motors, Mathematical Problems in Engineering (2013) 1–9, https://doi.org/10.1155/2013/715094.
2. A.Y. Alanis, E.N. Sanchez, A.G. Loukianov, Discrete-time adaptive backstepping nonlinear control via high-order

neural networks, IEEE Transactions on Neural Networks 18 (4) (2007) 1185–1195.

3. B.D.O. Anderson, J.B. Moore, Optimal Control: Linear Quadratic Methods, Prentice-Hall, Englewood Cliffs, NJ, USA, 1990.

4. K.J. Astrom, B. Wittenmark, Adaptive Control, 2nd ed., Pearson Education, Ma, USA, 1995.

5. M. Athans, P.L. Falb, Optimal Control: An Introduction to the Theory and Its Applications, McGraw Hill, New York, NY, USA, 1966.

6. P.J. Attar, E.H. Dowell, J.R. White, J.P. Thomas, Reduced order nonlinear system identification methodology, AIAA Journal 44 (8) (2006) 1895–1904.

7. H.T. Banks, S.C. Beeler, H.T. Tran, State estimation and tracking control of nonlinear dynamical systems, International Series of Numerical Mathematics 143 (2003) 1–24.

8. H.T. Banks, B.M. Lewis, H.T. Tan, Nonlinear feedback controllers and compensators: a state-dependent Riccati equation approach, Computational Optimization and Applications 37 (2) (2007) 177–218.

9. M.V. Basin, New Trends in Optimal Filtering and Control for Polynomial and Time-Delay Systems, Springer-Verlag, Berlin, Germany, 2008.

10. M.V. Basin, A.G. Loukianov, M. Hernandez-Gonzalez, Optimal controller for stochastic polynomial systems with state-dependent polynomial input, Circuits, Systems and Signal Processing 30 (6) (2011) 1463–1479.

11. K.H. Bellgardt, W. Kuhlmann, H.D. Meyer, K. Schugerl, M. Thoma, Application of an extended Kalman filter for state estimation of a yeast fermentation, IEE Proceedings D – Control Theory and Applications 133 (5) (1986) 226–234.

12. R.N. Bergman, Y.Z. Ider, C.R. Bowden, C. Cobelli, Quantitative estimation of insulin sensitivity, American Journal of Physiology: Endocrinology and Metabolism 236 (6) (1979) 667.

13. K. Bold, C. Edwards, J. Guckenheimer, S. Guharay, K. Hoffman, J. Hubbard, R. Oliva, W. Weckesser, The forced van der Pol equation II: canards in the reduced system, SIAM Journal on Applied Dynamical Systems 2 (4) (2003) 570–608.

14. B. Castillo-Toledo, S.D. Gennaro, A.G. Loukianov, J. Rivera, Discrete time sliding mode control with application to induction motors, Automatica 44 (12) (2008) 3036–3045.

15. F. Chen, M. Dunnigan, Comparative study of a sliding-mode observer and Kalman filters for full state estimation in an induction machine, IEE Proceedings Electric Power Applications 149 (1) (2002) 53–64.

16. J.R. Cloutier, C.N. D'Sousa, C.P. Mracek, Nonlinear regulation and nonlinear H_∞ control via the state-dependent Riccati equation technique: Part 1, theory, in: Proc. of the First Int. Conf. on Nonlinear Problems in Aviation and Aerospace, Daytoba Beach, FL, USA, 1996.

17. D.F. Coutinho, L.P.F.A. Pereira, A robust Luenberger-like observer for induction machines, in: Proceedings of the 31st Annual Conference of IEEE Industrial Electronics Society, IECON, Raleigh, NC, USA, 2005.

18. J. Kim, Efficient reduced-order system identification for linear systems with multiple inputs, AIAA Journal 43 (7) (2005) 1455–1464.

19. J.J. Craig, Introduction to Robotics: Mechanics and Control, Addison Wesley Longman, USA, 1989.

20. E.H. Dowell, K.C. Hall, M.C. Romanowski, Eigenmode analysis in unsteady aerodynamics reduced order models, Applied Mechanics Reviews 50 (1997) 371–386.

21. M.E. Fisher, A semiclosed-loop algorithm for the control of blood glucose levels in diabetics, Biomedical Engineering, IEEE Transactions on 38 (1) (1991) 57–61.

22. B. Friedland, Control System Design: An Introduction to State-Space Methods, McGraw-Hill, New York, NY, USA, 1986.

23. D. Funaro, Polynomial Approximation of Differential Equations, Springer, New York, NY, USA, 1992.

24. L.U. Gokdere, M.A. Simaan, A passivity-based method for induction motor control, IEEE Transactions on Industrial Electronics 44 (5) (1997) 688–695.

25. R. Gourdeau, Object-oriented programming for robotic manipulator simulation, IEEE Robotics and Automation 4 (3) (1997) 21–29.

26. R. Grover, P.Y.C. Hwang, Introduction to Random Signals and Applied Kalman Filtering, 2nd ed., John Wiley and Sons, New York, USA, 1992.

27. K.D. Hammett, C.D. Hall, D.B. Ridgely, Controllability issues in nonlinear state dependent Riccati equation control, Journal of Guidance, Control and Dynamics 21 (5) (1998) 767–773.

28. S. Haykin, Kalman Filtering and Neural Networks, Wiley, Upper Saddle River, NJ, USA, 2001.

29. R. Hovorka, V. Canonico, L.J. Chassin, U. Haueter, M. Massi-Benedetti, M. Orsini-Federici, T.R. Pieber, H. Schaller, L. Schaupp, T. Vering, M. Wilinska, Non-linear model predictive control of glucose concentration in subjects with type 1 diabetes, Physiological Measurement 25 (2004) 905–920.

30. H. Huang, G. Feng, J. Cao, Robust state estimation for uncertain neural networks with time-varying delay, IEEE Transactions on Neural Networks 19 (8) (2008) 1329–1339.

31. A. Isidori, Nonlinear Control Systems, Springer-Verlag, Berlin, Germany, 1995.

32. J.-N. Juang, Applied System Identification, Springer Science and Business Media, New York, NY, USA, 1994.

33. J.-N. Juang, R.S. Pappa, An eigensystem realization algorithm for modal parameter identification and model reduction, Journal of Guidance 8 (5) (1985) 620–627.

34. R.E. Kalman, A new approach to linear filtering and prediction problems, Journal of Basic Engineering 82 (1) (1960) 35–45.

35. P. Kaveh, Y.B. Shtessel, Blood glucose regulation using higher-order sliding mode control, International Journal of Robust and Nonlinear Control 18 (4–5) (2008) 557–569.

36. H.K. Khalil, Nonlinear Systems, 2nd ed., Prentice-Hall, Upper Saddle River, NJ, USA, 1996.

37. Y.H. Kim, F.L. Lewis, High-Level Feedback Control with Neural Networks, World Scientific, Singapore, 1998.

38. D.E. Kirk, Optimal Control Theory: An Introduction, Prentice-Hall, Englewood Cliffs, NJ, USA, 1970.

39. A.U. Levin, K.S. Narendra, Control of nonlinear dynamical systems using neural networks. II. Observability, identification, and control, IEEE Transactions on Neural Networks 7 (1) (1996) 30–42.

40. F.L. Lewis, V.L. Syrmos, Optimal Control, John Wiley & Sons, New York, NY, USA, 1995.

41. J. Li, Y. Zhong, Comparison of three Kalman filters for speed estimation of induction machines, in: Fortieth IAS Annual Meeting, Conference Record of the 2005 Industry Applications Conference, 2005, vol. 3, Kowloon, Hong Kong, China, 2005, pp. 1792–1797.

42. W. Lin, T. Shen, Robust passivity and control of nonlinear systems with structural uncertainty, in: Proceedings of the 36th Conference on Decision and Control, San Diego, CA, USA, 1997, pp. 2837–2842.

43. Y. Liu, Z. Wang, X. Liu, Design of exponential state estimators for neural networks with mixed time delays, Physics Letters A 364 (5) (2007) 104–412.

44. A.G. Loukianov, Nonlinear block control with sliding modes, Automation and Remote Control 57 (7) (1998) 916–933.

45. D.B. Madan, E. Seneta, Chebyshev polynomial approximations and characteristic function estimation, Journal of the Royal Statistical Society. Series B (Methodological) (1987) 163–169.

46. R. Marino, Adaptive observers for single output nonlinear systems, IEEE Transactions on Automatic Control 35 (9) (1990) 1054–1058.

47. K.S. Narendra, K. Parthasarathy, Identification and control of dynamical systems using neural networks, IEEE Transactions on Neural Networks 1 (1) (1990) 4–27.

48. O. Nelles, Nonlinear system identification, IOPScience 13 (4) (1987) 20–28.

49. N. Nikolaev, H. Iba, Adaptive Learning of Polynomial Networks: Genetic Programming, Backpropagation and Bayesian Methods, Springer Science and Business Media, New York, NY, USA, 2006.

50. R.D. Nowak, Nonlinear system identification, Circuits, Systems and Signal Processing 21 (1) (2002) 109–122.

51. S.-K. Oh, W. Pedrycz, B.-J. Park, Polynomial neural networks architecture: analysis and design, Computers and Electrical Engineering 29 (2003) 703–725.

52. F. Ornelas-Tellez, A.G. Loukianov, E.N. Sanchez, E.J. Bayro-Corrochano, Decentralized neural identification and control for uncertain nonlinear systems: application to planar robot, Journal of the Franklin Institute 347 (6) (2010) 1015–1034.

53. F. Ornelas-Tellez, J.J. Rico, R. Ruiz-Cruz, Optimal tracking for state-dependent coefficient factorized nonlinear systems, Asian Journal of Control (2013).

54. J.D. Pearson, Approximation methods in optimal control I. Sub-optimal control, Journal of Electronics and Control 13 (5) (1962) 453–465.

55. A.S. Poznyak, E.N. Sanchez, W. Yu, Differential Neural Networks for Robust Nonlinear Control, World Scientific, Singapore, 2001.

56. L.J. Ricalde, E.N. Sanchez, Inverse optimal nonlinear recurrent high order neural observer, in: Proceedings, 2005 IEEE International Joint Conference on Neural Networks, 2005, vol. 1, Montreal, Canada, 2005, pp. 361–365.

57. G.A. Rovithakis, M.A. Christodoulou, Adaptive Control with Recurrent High-Order Neural Networks, Springer-Verlag, Berlin, Germany, 2000.

58. E.N. Sanchez, A.Y. Alanis, Redes Neuronales: Conceptos Fundamentales y Aplicaciones a Control Automático, Pearson Education, España, 2006.

59. E.N. Sanchez, A.Y. Alanis, A.G. Loukianov, Discrete-Time High Order Neural Control, Springer-Verlag, Berlin, Germany, 2008.

60. E.N. Sanchez, A.Y. Alanis, J.J. Rico, Electric load demand and electricity prices ForecastingUsing higher order neural networks trained by Kalman filtering, in: M. Zhang (Ed.), Artificial Higher Order Neural Networks for Economics and Business, IGI Global, Hershey, PA, USA, 2009, pp. 295–313.

61. E.N. Sanchez, L.J. Ricalde, R. Langari, D. Shahmirzadi, Rollover prediction and control in heavy vehicles via recurrent neural networks, Intelligent Automation and Soft Computing 17 (1) (2011) 95–107.

62. S. Sastry, M. Bodson, Adaptive Control: Stability, Convergence and Robustness, Courier Dover Publications, 2011.

63. S. Singhal, L. Wu, Training multilayer perceptrons with the extended Kalman algorithm, in: D.S. Touretzky (Ed.), Advances in Neural Information Processing Systems 1, Morgan Kaufmann Publishers Inc., San Francisco, CA, USA, 1989, pp. 133–140.

64. Y. Song, J.W. Grizzle, The extended Kalman filter as local asymptotic observer for discrete-time nonlinear systems, Journal of Mathematical Systems 5 (1) (1995) 59–78.

65. R.F. Stengel, Optimal Control and Estimation, revised ed., Dover Publications, Inc., New York, NY, USA, 1994 (Originally published by John Wiley and Sons, New York, 1986).

66. V. Sundarapandian, Reduced order observer design for nonlinear systems, Applied Mathematics Letters 19 (2006) 936–941.

67. D. Tang, D. Kholodar, J.-N. Juang, E.H. Dowell, System identification and proper orthogonal decomposition method applied to unsteady aerodynamics, AIAA Journal 39 (8) (2001) 1569–1576.

68. V. Utkin, Sliding mode control design principle and applications to electrical drives, IEEE Transactions on Industrial Electronics 40 (1993) 23–36.

69. V. Utkin, J. Guldner, J. Shi, Sliding Mode Control in Electromechanical Systems, Taylor and Francis, Philadelphia, USA, 1999.

70. M. Vidyasagar, Nonlinear Systems Analysis, 2nd ed., Prentice-Hall, Englewood Cliffs, NJ, USA, 1993.

71. B.L. Walcott, S. Zak, State observation of nonlinear uncertain dynamical system, IEEE Transactions on Automatic Control 32 (1987) 166–170.

72. Z. Wang, D.W.C. Ho, X. Liu, State estimation for delayed neural networks, IEEE Transactions on Neural Networks 16 (1) (2005) 279–284.

73. A. Weiss, I. Kolmanovsky, D.S. Bernstein, Forward-integration Riccati-based output-feedback control of linear time-varying systems, in: American Control Conference (ACC), 2012, IEEE, 2012, pp. 6708–6714.

CHAPTER 4

Continuous-Time Decentralized Neural Control of a Quadrotor UAV

FRANCISCO JURADO, DSC • SERGIO LOPEZ, MSC

4.1 INTRODUCTION

Research on artificial neural networks (ANNs) enjoys great interest mainly due to its capabilities when approximating any continuous function. Another reason for studying ANNs is because they have shown very good adaptive and learning capabilities in the presence of external disturbances and uncertainties in modeling. With the use of ANNs, control algorithms have been developed to be robust to uncertainties and modeling errors. In recent years, the use of recurrent high-order neural networks (RHONNs) for identification and control has increased [11,10,1,37,2,26,25]. The use of RHONNs allows to propose simple neural identification schemes, i.e., structures of a single layer, unlike some structures such as the one proposed in [12]. Likewise, the implementation of an online training algorithm allows to take full advantage of the capabilities of approximation of the ANN in order to identify, very accurately, the dynamic behavior of a system, including nonlinearities due to friction.

The best well-known training approach for a recurrent neural network (RNN) is the backpropagation through time learning algorithm. However, it is a first-order gradient descent method and hence its learning speed could be very slow. Since the late 1980s, the extended Kalman filter (EKF)-based algorithms have been introduced to train ANNs, showing good results when improving learning convergence. EKF training of ANNs, both feedforward and recurrent ones, has proven to be reliable and practical for many applications, such as those discussed in [9]. It is known that for many nonlinear systems, it is often a challenge to obtain accurate and reliable mathematical models, regarding their physically complex structures and hidden parameters. Therefore, system identification becomes an interesting problem and even necessary before system control can be considered, not only for understanding and predicting the behavior of the system, but also to obtain an effective control law. The neural identification problem consists of selecting an appropriate neural identification model, to then adjust its parameters according to

an adaptive law, such that the response of the neural identifier model to an input signal, or a class of input signals, approximates the response of the real system for the same input [38].

The fast advance in control technology offers new ways for implementing neural control algorithms within the approach of a centralized control design. Nevertheless, there is a great challenge to obtain an efficient control for centralized systems due to the highly nonlinear complex dynamics, the presence of strong interconnections, parameters that are difficult to determine, and unmodeled dynamics. Considering only the most important terms, the mathematical model obtained could need a control algorithm with a great number of mathematical operations, becoming unfeasible for real-time implementation. An alternative approach has been developed considering a global system as a set of interconnected subsystems, for which it is possible to design independent controllers, considering only local variables to each subsystem. This approach is the so-called *decentralized control* scheme [22]. The decentralized control approach has been applied to different research fields, like robotics, mainly in cooperative multiple mobile robots and robot manipulators. For robot manipulators each joint and its respective link is considered as a subsystem in order to develop a local controller just considering the local angular position and angular velocity measurements, compensating the interconnection effects and assuming them as disturbances. The resulting controllers in most cases are easy to implement in real-time [24,30]. Focusing on neural control of robot manipulators, interesting works have been reported in the literature, such as [33], where a decentralized control scheme for a robot manipulator was developed decoupling the dynamic model in a set of linear subsystems with uncertainties showing simulation results for a two-joint robot. In [39], a decentralized control for a robot manipulator was reported, which was based on the estimation of each dynamic's joint using feedforward neural networks (FFNN). In the same way, in [35] a discrete-time decentralized neural identification and control scheme was presented, which

Artificial Neural Networks for Engineering Applications. https://doi.org/10.1016/B978-0-12-818247-5.00013-7

was developed using a RHONN trained via an EKF algorithm. The discrete-time control law proposed was based on a block control scheme (BCS) and sliding mode techniques the performance of which was validated via simulation and later implemented in real-time on a two–degree of freedom (DOF) planar robot.

In the recent literature about the use of backstepping control schemes, numerous approaches have been proposed for the design of neural controllers. In [13], an adaptive backstepping ANN control approach was extended for a class of large-scale nonlinear output feedback systems with completely unknown and mismatched interconnections. In [20], a discrete-time decentralized neural control scheme for identification and trajectory tracking of a 2-DOF robot manipulator was presented. A RHONN structure was proposed to identify the plant model and a discrete-time control law was derived combining discrete-time BCS and sliding mode techniques. In [19], a high-order neural network (HONN) was used to approximate a decentralized control law designed using the backstepping approach as applied for a block strict feedback form (BSFF). The ANN learning was performed online by Kalman filtering. Besides, an application in real-time on a redundant robot manipulator of five DOFs was presented in [18].

In [8], a time-varying learning algorithm based on the EKF for a RHONN and a decentralized discrete-time BCS were proposed in order to assess the trajectory tracking of a 5-DOF robot manipulator. In [21], a discrete-time HONN was proposed to identify, in a decentralized form, the dynamic behavior of a Quanser 2-DOF helicopter model. Also, the neural backstepping and sliding mode BCS techniques were considered in order to deal with the decentralized control problem for the output trajectory tracking to both pitch and yaw positions. Many works on continuous-time HONN structures for identification, using the filtered error (FE) learning law [23], and control of robot manipulators with results in real-time have been reported. A continuous-time decentralized control strategy using a recursive method and based on an RNN identifier with a BCS for identification and control was proposed in [40]. The ANN learning law was performed using the FE technique. This control strategy was tested via simulation on a 2-DOF vertical robot manipulator model, for which terms due to friction were not considered. An adaptive tracking controller algorithm in continuous time for nonlinear multiple input–multiple output (MIMO) systems was presented in [3]. This algorithm included the combination of a RHONN with high-order sliding mode techniques. The ANN was used to identify the dynamics for the plant where the FE algorithm was used to train the neural identifier. A decentralized high-order sliding mode scheme in twisting modality was used to design reduced chattering–independent controllers to develop the trajectory tracking for a 3-DOF robot arm via simulation. In [42], a continuous-time decentralized neural control scheme for trajectory tracking of a 2-DOF direct drive vertical robotic arm powered by industrial servomotors was presented. A decentralized RHONN structure was proposed to identify online, in a series-parallel configuration and using the FE learning law, the dynamics of the plant. Based on the RHONN subsystems, a local neural controller was derived via the backstepping approach. The performance of the decentralized RHONN controller was validated via experimental results.

A key aspect of the design of an unmanned aerial vehicle (UAV) is an autonomous or remotely programmable mission planner which is broadly responsible for establishing the route from the point of take-off, via a set of prespecified waypoints, to the point of landing, a navigation system whose primary role is to determine the location and orientation of the vehicle at any given time, a guidance system which is responsible for determining the desired flight path by specifying the position and orientation commands for the vehicle from one waypoint to the next, and a flight controller which is responsible for ensuring that the vehicle tracks the guidance commands [43]. The design of a flight control system is essentially an exercise in designing a feedback controller that meets all of the requirements over a specific mission. In particular, a UAV is required to perform risky or tedious operations under extreme operational conditions. Moreover, flight control systems for a UAV are considered to be mission-critical, as a flight control system failure could result in either an unsuccessful mission or the loss of the UAV. While linear or gain-scheduled linear controllers may be typically adequate for these operations, it is often found that the UAV dynamics is nonlinear and that the linear controllers will not meet the performance requirements even when the gains are scheduled. This is due to the adverse coupling between the longitudinal and lateral dynamics that necessitates additional feedback. Furthermore, the uncertainty in the parameters of the mathematical model of the UAV and the environmental uncertainties are significant, and this can be another factor in making the design of long-endurance, high-integrity autonomous flight controllers based on linear models, which are extremely challenging. To meet these increasing demands on stability, performance, and reliability of UAVs, nonlinear and adaptive control techniques are often used [29,41,43]. These techniques are actively being researched to handle nonlinear aerodynamic and kinematic effects, actuator saturations and rate limitations, modeling and parameter uncertainty, and parameter-varying dynamics.

A typical control design starts with modeling, which is basically a procedure of constructing a mathematical description for the physical system to be controlled. This selected model needs to reflect main features of the physical process. Accurate models are not always better. They may require unnecessarily complex control design and demand excessive computations. From a control point of view, the key in modeling is to capture the essential effects in the system dynamics within an operating range of interest. In addition, a good model should also provide some characterization of the system uncertainties [29]. In this sense, the use of ANNs for the control of a quadrotor UAV can be a good choice. In [44], a nonlinear control system for a quadrotor based on a combination of state-dependent Riccati equations and ANNs was developed. The performance of this control system was validated via numerical simulation. An adaptive neural network control to stabilize a quadrotor against modeling errors and wind disturbance was proposed in [34]. The method was compared to both dead-zone and e-modification adaptive techniques via numerical simulation. A nonlinear adaptive controller for a quadrotor using the backstepping technique mixed with ANNs was proposed in [31]. The backstepping approach was used to achieve tracking of the desired translation positions and yaw angle while maintaining the stability of pitch and roll angles simultaneously. The knowledge of all physical parameters and the exact model of the quadrotor were not required for the controller. Online adaptation of the ANNs and some parameters was used to compensate some unmodeled dynamics, including aerodynamic effects. The feasibility of the control scheme was demonstrated through simulation results. In [14], an ANN-based output feedback controller for a quadrotor UAV was proposed. The ANNs were utilized in an observer to generate virtual and actual control inputs where the ANNs learn the nonlinear dynamics of the quadrotor UAV online, including uncertain nonlinear terms like aerodynamic friction and blade flapping. In [15] a nonlinear controller for a quadrotor UAV using ANNs and output feedback was proposed. An ANN was introduced to learn the dynamics of the UAV online, including uncertain nonlinear terms like aerodynamic friction and blade flapping. An ANN virtual control input scheme was proposed allowing all six DOFs of the UAV to be controlled using only four control inputs. An ANN observer was introduced to estimate the translational

and angular velocities. An output feedback control law was developed in which only the position and attitude of the quadrotor UAV were considered as measurable. Simulation results demonstrated the effectiveness of the output feedback control scheme. An adaptive neural control scheme based on an observer applied to a quadrotor was proposed in [7]. Two parallel feedforward ANNs for each subsystem of the quadrotor were used. The first one estimates online the equivalent control term and the second one generates a corrective term for the observer. The performance of the adaptive neural control scheme was validated via simulation results. In [17], the effectiveness of the direct inverse control technique using an ANN to learn and cancel out the hover dynamics of the quadrotor UAV under various environmental conditions during a hover mode was investigated. In [45] an adaptive inverse model control method was proposed for a quadrotor UAV. Two backpropagation networks were used in order to achieve the model identification and control. Both offline training and online learning were used, ensuring fast learning and robustness. The performance of the proposal was validated via simulation. In [5], an ANN-based controller for a quadrotor UAV was proposed where an indirect model reference adaptive control scheme was used to show trajectory tracking in the presence of dynamically modeled thrust and drag coefficients. In [4], a RHONN trained with the EKF was used to identify the dynamics from the motors of a quadrotor UAV where inverse optimal control was employed for stabilization of propeller speeds. The feasibility of the proposed control strategy was validated via simulation results.

In this work, we describe in detail the design of a continuous-time decentralized RHONN controller for a quadrotor UAV, previously presented in [27], where it was taken as baseline controller. Along the same line as in [42], from the strict-feedback structure proposed for the RHONN, a local neural controller is derived using the backstepping design methodology. The training for the RHONN is performed online, in a series-parallel configuration, using the FE algorithm. The effectiveness of our decentralized neural control scheme is validated via numerical simulation results.

4.2 FUNDAMENTALS

4.2.1 Recurrent High-Order Neural Network (RHONN)

Consider the RHONN model [38] where the state of each neuron is governed by a differential equation of the form

$$\dot{x}_j^i = -a_j^i x_j^i + b_j^i \sum_{k=1}^{L} w_{jk}^i \prod_{j \in I_k} y_j{}^{d_j(k)}, \qquad (4.1)$$

where x_j^i is the state of the ith neuron, a_j^i, b_j^i are positive real constants, w_{jk}^i is the kth adjustable synaptic weight connecting the jth state to the ith neuron, L represents the total number of weights used to identify the plant behavior, and y_j is the activation function for each one of the connections. Each y_j is either an external input or the state of a neuron through a sigmoidal function, i.e., $y_j = s(x_j^i)$, where $s(\cdot)$ is a sigmoidal nonlinearity. In a recurrent second-order neural network, the total input to the neuron is not only a linear combination of the components y_j, it may also be the product of two to more elements represented by triplets $y_1 y_2 y_3$, quadruplets, etc; $\{I_1, I_2, \ldots, I_L\}$ is a collection of L nonordered subsets of $\{1, 2, \ldots, i + j\}$ and $d_j(k)$ are nonnegative integers. This class of neural networks forms a RHONN.

The input vector for each neuron is given by

$$y = \begin{pmatrix} y_1 \\ \vdots \\ y_i \\ y_{i+1} \\ \vdots \\ y_{i+j} \end{pmatrix} = \begin{pmatrix} s(x^1) \\ \vdots \\ s(x^i) \\ u^1 \\ \vdots \\ u^j \end{pmatrix}, \qquad (4.2)$$

where $u = [u^1\, u^2\, \cdots\, u^j]^\top$ is the vector of external control inputs to the network; the superscript $\top$ denotes the transposed vector.

Introducing an L-dimensional vector z, defined as

$$z = \begin{pmatrix} z_1 \\ z_2 \\ \vdots \\ z_L \end{pmatrix} = \begin{pmatrix} \prod_{j \in I_1} y_j{}^{d_j(1)} \\ \prod_{j \in I_2} y_j{}^{d_j(2)} \\ \vdots \\ \prod_{j \in I_L} y_j{}^{d_j(L)} \end{pmatrix}, \qquad (4.3)$$

the RHONN model (4.1) can be rewritten as

$$\dot{x}_j^i = -a_j^i x_j^i + b_j^i \sum_{k=1}^{L} w_{jk}^i z_{jk}^i. \qquad (4.4)$$

We define the adjustable parameter vector as $w_{jk}^i = b_j^i [w_{j1}^i\, w_{j2}^i\, \cdots\, w_{jL}^i]^\top$, so (4.4) becomes

$$\dot{x}_j^i = -a_j^i x_j^i + (w_{jk}^i)^\top z_{jk}^i, \qquad (4.5)$$

where the vectors w_{jk}^i represent the adjustable weights of the network, while the coefficients a_j^i for $i = 1, 2,$

$\ldots, n$ are part of the underlying network architecture and are fixed during training.

In order to guarantee that each neuron x_j^i is bounded input–bounded output (BIBO)-stable, we assume that $a_j^i > 0$. The dynamic behavior of the overall network is described by expressing (4.5) in vector notation as follows:

$$\dot{x} = Ax + W^\top \psi, \qquad (4.6)$$

where $x = [x_j^1, x_j^2, \ldots, x_j^n]^\top \in \mathbb{R}^n$, $W = [w_j^1, w_j^2, \ldots, w_j^n]^\top \in \mathbb{R}^{L \times n}$, and $A = \text{diag}[-a_j^1, -a_j^2, \ldots, -a_j^n] \in \mathbb{R}^{n \times n}$ is a diagonal matrix. Since $a_j^i > 0, \forall i = 1, 2, \ldots, n$, A is Hurwitz.

4.2.2 Approximation Properties of the RHONN

In the following, the problem of approximating a general nonlinear dynamical system by a RHONN is described. The input–output behavior of the system to be approximated is given by

$$\dot{\chi}_j^i = F(\chi_j^i, u^i), \qquad (4.7)$$

where $u^i \in \mathbb{R}^i$ is the input to the system, $\chi_j^i \in \mathbb{R}^j$ is the state of the system, and $F : \mathbb{R}^{i+j} \mapsto \mathbb{R}^j$ is a smooth vector field defined on a compact set $\mathcal{Y} \subset \mathbb{R}^{i+j}$, where i and j are constants. The approximation problem consists of determining, by allowing enough high-order connections, if there exist weights w_{jk}^i such that (4.5) approximates the input–output behavior of an arbitrary dynamical system of the form (4.7). Assume that $F(\cdot)$ is continuous and satisfies a local Lipschitz condition such that (4.7) has a unique solution and $(\chi_j^i(t), u^i(t)) \in \mathcal{Y}$ for all t in some time interval $J_T = \{t : 0 \le t \le T\}$, where J_T represents the time period over which the approximation is performed. From (4.1), (4.2) it is clear that $z(x, u)$ is in the standard polynomial expansion with the exception that each component of the vector x is preprocessed by a sigmoid function $s(\cdot)$. Based on the above assumptions, the next theorem, which is strictly an existence result and does not provide any constructive method in order to obtain the optimal weights w^{*i}_{jk}, proves that if a sufficiently large number of high order terms is allowed in (4.5), then it is possible to approximate any dynamical system to any degree of accuracy.

Theorem 4.1. *Suppose that the system (4.7) and the RHONN model (4.5) are initially at the same state $\chi_j^i(0) = x_j^i(0)$. Then, for any $\varepsilon > 0$ and any finite $T > 0$ there exist an integer L and a vector $w^{*i}_{jk} \in \mathbb{R}^L$ such that the state*

$x_j^i(t)$ of the RHONN model (4.5), with L high-order connections and weight values $w_{jk}^i = w_{jk}^{*i}$, satisfies

$$sup_{0 \leq t \leq T}|x_j^i(t) - \chi_j^i(t)| \leq \varepsilon.$$

Proof. See [38]. ☐

4.2.3 Filtered Error Training Algorithm

Under the assumption that the unknown system is exactly modeled by a RHONN architecture of the form (4.5), the weight adjustment law and the FE training algorithm for this RHONN are next summarized. Based on the assumptions of no modeling error, there exist unknown weight vectors w_{jk}^{*i} such that each state χ_j^i of the unknown dynamical system (4.7) satisfies

$$\dot{\chi}_j^i = -a_j^i \chi_j^i + (w_{jk}^{*i})^\top z_{jk}^i(\chi_j^i, u^i), \quad \chi_j^i(0) = \chi_{j0}^i, \tag{4.8}$$

where χ_{j0}^i is the initial state of the system. As is standard in systems identification procedures, here it is assumed that the input $u^i(t)$ and the state $\chi_j^i(t)$ remain bounded for all $t \geq 0$. Based on the definition for $z_{jk}^i(\chi_j^i, u^i)$ given by (4.3), this implies that $z_{jk}^i(\chi_j^i, u^i)$ is also bounded. In the sequel, unless there might exist confusion, the arguments of the vector field z will be omitted. Next, the approach for estimating the unknown parameters w_{jk}^{*i} of the RHONN model (4.8) is described.

Considering (4.8) as the differential equation describing the dynamics of the unknown system, the identifier structure is chosen with the same form as in (4.5), where w_{jk}^i is the estimate of the unknown weight vector w_{jk}^{*i}. From (4.5) and (4.8), the identification error $\xi_j^i = x_j^i - \chi_j^i$ satisfies

$$\dot{\xi}_j^i = \dot{x}_j^i - \dot{\chi}_j^i, \tag{4.9}$$
$$= -a_j^i(x_j^i - \chi_j^i) + ((w_{jk}^i)^\top - (w_{jk}^{*i})^\top)z_{jk}^i, \tag{4.10}$$

which can be rewritten as

$$\dot{\xi}_j^i = -a_j^i \xi_j^i + (\tilde{w}_{jk}^i)^\top z_{jk}^i, \tag{4.11}$$

where $(\tilde{w}_{jk}^i)^\top = (w_{jk}^i)^\top - (w_{jk}^{*i})^\top$ denotes the parametric error [42]. The weights w_{jk}^i are adjusted according to the learning law, i.e.,

$$\dot{w}_{jk}^i = -\Gamma_{jk}^i z_{jk}^i \xi_j^i, \tag{4.12}$$

where the adaptive gain $\Gamma_{jk}^i \in \mathbb{R}^{L \times L}$ is a positive definite matrix. Stability and convergence properties for the weight adjustment law given above are analyzed in [23]. The following theorem establishes that this identification scheme has convergence properties with the gradient method for adjusting the weights.

Theorem 4.2. *Consider the filtered error RHONN model given by (4.11), whose weights are adjusted according to (4.12). Then:*
1. *$\xi_j^i, \tilde{w}_{jk}^i \in \mathcal{L}_\infty$ (i.e., ξ_j^i and $\tilde{w}_{jk}^i$ are uniformly bounded);*
2. *$lim_{t \to \infty} \xi_j^i(t) = 0$.*

Proof. See [28]. ☐

4.3 NEURAL BACKSTEPPING CONTROLLER DESIGN

In this work a miniature four-rotor helicopter is considered, having two of these rotors rotating clockwise, and two rotating counterclockwise. Each rotor consists of a direct current brushless motor with propeller. Forward motion is accomplished by increasing the speed of the rear rotor while simultaneously reducing the speed of the forward rotor with the same amount. Backward, leftward, and rightward motion can be accomplished in the same way. Finally, yaw motion can be performed accelerating the two clockwise turning rotors, while decelerating the counterclockwise ones.

The equations describing the attitude and position of a quadrotor are those of a rotating rigid body with six DOFs [16]. They can be separated into kinematic and dynamic equations [32]. The dynamic equations can be obtained around the center of mass, i.e.,

$$F_{ext} = m_q \dot{V}_b + \omega \times m_q V_b, \tag{4.13}$$
$$T_{ext} = J\dot{\omega} + \omega \times J\omega, \tag{4.14}$$

where m_q denotes the quadrotor mass, V_b is the velocity in the body frame, ω is the angular rate of the quadrotor, $J = diag\{J_x, J_y, J_z\}$ is the inertia matrix, and the external force $F_{ext} \in \mathbb{R}^3$ takes into account the quadrotor weight, the total thrust, and the aerodynamic force, whereas the external torque $T_{ext} \in \mathbb{R}^3$ considers the difference of thrust and torque exerted by the two pairs of rotors as well as the aerodynamic moment vector.

The equations of motion for a quadrotor, assuming low speeds, are given by [6]

$$\ddot{x} = (c\phi c\psi s\theta + s\phi s\psi)\frac{F}{m_q}, \tag{4.15}$$

$$\ddot{y} = (c\phi s\theta s\psi - c\psi s\phi)\frac{F}{m_q}, \tag{4.16}$$

$$\ddot{z} = -g + (c\theta c\phi)\frac{F}{m_q}, \tag{4.17}$$

$$\ddot{\theta} = \left(\frac{J_z - J_x}{J_y}\right)\dot{\phi}\dot{\psi} + \frac{J_r}{J_y}\varpi_r\dot{\phi} + \frac{l}{J_y}\tau_y, \tag{4.18}$$

$$\ddot{\phi} = \left(\frac{J_y - J_z}{J_x}\right)\dot{\theta}\dot{\psi} - \frac{J_r}{J_x}\varpi_r\dot{\theta} + \frac{l}{J_x}\tau_x, \tag{4.19}$$

$$\ddot{\psi} = \left(\frac{J_x - J_y}{J_z}\right)\dot{\theta}\dot{\phi} + \frac{1}{J_z}\tau_z, \tag{4.20}$$

where x, y, and z are the coordinates to the center of mass in the inertial frame, θ, ϕ, and ψ are Euler angles representing pitch, roll, and yaw, respectively, $s(\cdot)$, $c(\cdot)$ denote the $\sin(\cdot)$ and $\cos(\cdot)$ functions, J_x, J_y, and J_z are moments of inertia in the direction of the three-dimensional Cartesian coordinates, l represents the distance between the rotors with respect to the center of mass (center of gravity) of the quadrotor, $\varpi_r = -\varpi_1 + \varpi_2 - \varpi_3 + \varpi_4$ is the sum of angular velocity of the rotors, and J_r represents the inertia of the rotating rotors, which is considered the same parameter for all four motors.

Consider that the system (4.15)–(4.20) can be divided into N subsystems of the form [40]

$$\dot{\chi}_2^i = f^i(\chi_1^i, \chi_2^i, u^i) + v^i(\chi, \dot{\chi}, U), \tag{4.21}$$

where $i = 1, 2, ..., N$, $f^i(\cdot) \in \mathbb{R}$ only depends on local variables, $v^i(\cdot) \in \mathbb{R}$ represents interconnection effects, $\chi_1^i, \chi_2^i \in \mathbb{R}$ represent the position and velocity from the ith subsystem, respectively, $u^i \in \mathbb{R}$ is the input for each subsystem, $\chi \in \mathbb{R}^n$ is the state vector of the system, and $U \in \mathbb{R}^m$ is the input vector of the system. For this purpose, the governing equations can be expressed in the state equation through the selection of a set of state variables given by

$$
\begin{aligned}
&x_1 = \phi, \quad x_3 = \theta, \quad x_5 = \psi, \quad x_7 = z, \quad x_9 = x, \quad x_{11} = y,\\
&x_2 = \dot{\phi}, \quad x_4 = \dot{\theta}, \quad x_6 = \dot{\psi}, \quad x_8 = \dot{z}, \quad x_{10} = \dot{x}, \quad x_{12} = \dot{y}.
\end{aligned} \tag{4.22}
$$

Then, the system dynamics can be divided in a decentralized way as

$$S_x \begin{cases} \dot{x}_9 = x_{10}, \\ \dot{x}_{10} = (cx_1 sx_3 cx_5 + sx_1 sx_5)\dfrac{F}{m_q}, \end{cases} \tag{4.23}$$

$$S_y \begin{cases} \dot{x}_{11} = x_{12}, \\ \dot{x}_{12} = (cx_1 sx_3 sx_5 - sx_1 cx_5)\dfrac{F}{m_q}, \end{cases} \tag{4.24}$$

$$S_z \begin{cases} \dot{x}_7 = x_8, \\ \dot{x}_8 = -g + (cx_1 cx_3)\dfrac{F}{m_q}, \end{cases} \tag{4.25}$$

$$S_\theta \begin{cases} \dot{x}_3 = x_4, \\ \dot{x}_4 = \left(\dfrac{J_z - J_x}{J_y}\right)x_2 x_6 + \dfrac{J_r}{J_y}\varpi_r x_2 + \dfrac{l}{J_y}\tau_y, \end{cases} \tag{4.26}$$

$$S_\phi \begin{cases} \dot{x}_1 = x_2, \\ \dot{x}_2 = \left(\dfrac{J_y - J_z}{J_x}\right)x_4 x_6 - \dfrac{J_r}{J_x}\varpi_r x_4 + \dfrac{l}{J_x}\tau_x, \end{cases} \tag{4.27}$$

$$S_\psi \begin{cases} \dot{x}_5 = x_6, \\ \dot{x}_6 = \left(\dfrac{J_x - J_y}{J_z}\right)x_2 x_4 + \dfrac{1}{J_z}\tau_z. \end{cases} \tag{4.28}$$

Thus, the neural identification problem for the whole system can be simplified through the use of six second-order subsystems, namely one second-order subsystem for each coordinate.

In this chapter, a decentralized RHONN trained via the FE algorithm (4.11) with weights adjustment law (4.12) is proposed for identification and control of a quadrotor UAV. From (4.5), the decentralized RHONN model is given as follows:

$$
\begin{aligned}
\dot{x}_1^i &= -a_1^i x_1^i + w_{11}^i z_{11}^i(\chi_1^i) + x_2^i, \tag{4.29}\\
\dot{x}_2^i &= -a_2^i x_2^i + w_{21}^i z_{21}^i(\chi_1^i) + w_{22}^i z_{22}^i(\chi_2^i)\\
&\quad + w_{23}^i z_{23}^i(\chi_1^i, \chi_2^i) + u^i,
\end{aligned}
$$

with $k = 1, 2, 3$ for the $w_{jk}^i z_{jk}^i(\cdot)$ term, where $j = 1, 2$ is for the number of states of the ith RHONN model; χ_1^i represents the measurable local translational (angular) position and χ_2^i is for the calculated local translational (angular) velocity; u^i is the control input. The decentralized RHONN model (4.29) can be expressed in compact form as

$$
\begin{aligned}
\dot{x}_1^i &= -a_1^i x_1^i + (w_1^i)^\top z_1^i + x_2^i, \tag{4.30}\\
\dot{x}_2^i &= -a_2^i x_2^i + (w_2^i)^\top z_2^i + u^i.
\end{aligned}
$$

It must be noticed that this decentralized neural scheme is in the form of a strict-feedback system; then, the use of the backstepping approach results to be suitable for the design of the neural controller. For each ith subsystem the identification error between the neural identifier and the coordinate is defined as $\xi_1^i = x_1^i - \chi_1^i$ for the translational (angular) position and $\xi_2^i = x_2^i - \chi_2^i$ for the translational (angular) velocity. To update online the synaptic weights, the adaptive learning laws are given by

$$
\begin{aligned}
\dot{w}_{11}^i &= -\gamma_{11}^i z_1^i \xi_1^i,\\
\dot{w}_{21}^i &= -\gamma_{21}^i z_1^i \xi_2^i,
\end{aligned}
$$

$$\dot{w}^i_{22} = -\gamma^i_{22} z^i_2 \xi^i_2,$$
$$\dot{w}^i_{23} = -\gamma^i_{23} z^i_3 \xi^i_2,$$

with $\gamma^i_{jk} > 0$ as the adaptive gain and

$$z^i_1 = \tanh(\chi^i_1),$$
$$z^i_2 = \tanh(\chi^i_2),$$
$$z^i_3 = \tanh(\chi^i_1)\tanh(\chi^i_2).$$

Next, our objective is to design a feedback control law u^i to force the system output to follow a desired trajectory. The decentralized neural control scheme is based on the following.

Denoting $\xi^i = |\chi^i - x^i|$ as the identification error and the trajectory tracking error between the states of the neural network for position and the desired trajectory as $\epsilon^i = |x^i - \chi^i_d|$, the output tracking error is rewritten as

$$\tilde{\chi}^i = \xi^i + \epsilon^i. \tag{4.31}$$

Consequently, the error dynamics is given as

$$\dot{\tilde{\chi}}^i = \dot{\xi}^i + \dot{\epsilon}^i. \tag{4.32}$$

Proposing the Lyapunov-like function

$$V^i(\xi^i, \tilde{w}^i) = \frac{1}{2}\sum_{j=1}^n \left((\xi^i_j)^2 + (\tilde{w}^i_j)^\top (\gamma^i_j)^{-1}(\tilde{w}^i_j)\right),$$

which stabilizes ξ^i, the time derivative $\dot{V}^i(\cdot)$ of $V^i(\cdot)$ is given by

$$\dot{V}^i(\xi^i, \tilde{w}^i) = \sum_{j=1}^n \left(\xi^i_j \dot{\xi}^i_j + (\tilde{w}^i_j)^\top (\gamma^i_j)^{-1}\dot{w}^i_j\right),$$

which evaluated along the trajectories (4.11) and (4.12) becomes

$$\dot{V}^i(\xi^i, \tilde{w}^i) = \sum_{j=1}^n -a^i_j (\xi^i_j)^2 \le 0. \tag{4.33}$$

Now, from the proposed neural structure (4.29), we proceed with the design of the controller that stabilizes ϵ^i. To this end, defining the trajectory tracking errors of the RHONN as

$$\epsilon^i_1 = x^i_1 - \chi^i_{1d},$$
$$\epsilon^i_2 = x^i_2 - \chi^i_{2d}, \tag{4.34}$$

the error dynamics are given by

$$\dot{\epsilon}^i_1 = -a^i_1 \epsilon^i_1 - a^i_1 \chi^i_{1d} + (w^i_1)^\top z^i_1 + \epsilon^i_2, \tag{4.35}$$
$$\dot{\epsilon}^i_2 = -a^i_2 \epsilon^i_2 - a^i_2 \chi^i_{2d} + (w^i_2)^\top z^i_2 + u^i - \dot{\chi}^i_{2d}. \tag{4.36}$$

From (4.35), considering ϵ^i_2 as virtual control input, we proceed with the design of a control $\epsilon^i_2 = \alpha^i_2$ in order to stabilize the origin $\epsilon^i_1 = 0$. Therefore, choosing

$$\epsilon^i_2 = \alpha^i_2 = a^i_1 \chi^i_{1d} - (w^i_1)^\top z^i_1 - c^i_1 \epsilon^i_1 \tag{4.37}$$

the nonlinear terms are canceled. Consequently,

$$\dot{\epsilon}^i_1 = -a^i_1 \epsilon^i_1 - c^i_1 \epsilon^i_1. \tag{4.38}$$

Now, proposing the augmented Lyapunov-like function

$$V^i_1(\epsilon^i_1, \xi^i, \tilde{w}^i) = V^i(\xi^i, \tilde{w}^i) + \frac{1}{2}(\epsilon^i_1)^2, \tag{4.39}$$

the time derivative of (4.39) is given by

$$\dot{V}^i_1(\epsilon^i_1, \xi^i, \tilde{w}^i) = \dot{V}^i(\xi^i, \tilde{w}^i) + \epsilon^i_1 \dot{\epsilon}^i_1. \tag{4.40}$$

From (4.38), (4.40) can be written as

$$\dot{V}^i_1(\epsilon^i_1, \xi^i, \tilde{w}^i) = \dot{V}^i(\xi^i, \tilde{w}^i) + \epsilon^i_1(-a^i_1 \epsilon^i_1 - c^i_1 \epsilon^i_1), \tag{4.41}$$

or equivalently,

$$\dot{V}^i_1(\epsilon^i_1, \xi^i, \tilde{w}^i) = \dot{V}^i(\xi^i, \tilde{w}^i) - (a^i_1 + c^i_1)(\epsilon^i_1)^2,$$
$$\forall \epsilon^i_1, \xi^i, \tilde{w}^i \in \mathbb{R}. \tag{4.42}$$

Thus, the origin of $\dot{\epsilon}^i_1$ is globally exponentially stable.

From the backstepping approach, applying the variable change

$$\zeta^i_2 = \epsilon^i_2 - \alpha^i_2 = \epsilon^i_2 - a^i_1 \chi^i_{1d} + (w^i_1)^\top z^i_1 + c^i_1 \epsilon^i_1, \tag{4.43}$$

we have

$$\epsilon^i_2 = \zeta^i_2 + \alpha^i_2 = \zeta^i_2 + a^i_1 \chi^i_{1d} - (w^i_1)^\top z^i_1 - c^i_1 \epsilon^i_1. \tag{4.44}$$

Substituting (4.44) in (4.35) yields

$$\dot{\epsilon}^i_1 = -a^i_1 \epsilon^i_1 - c^i_1 \epsilon^i_1 + \zeta^i_2. \tag{4.45}$$

Continuing with the backstepping approach, the time derivative of (4.37) results in

$$\dot{\alpha}^i_2 = a^i_1 \chi^i_{2d} - (\dot{w}^i_1)^\top z^i_1 - (\dot{z}^i_1)^\top w^i_1 - c^i_1 \dot{\epsilon}^i_1. \tag{4.46}$$

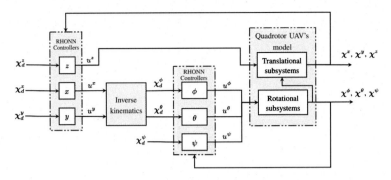

FIG. 4.1 Decentralized RHONN control scheme for the quadrotor UAV.

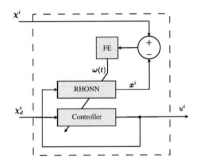

FIG. 4.2 RHONN controller.

From (4.45), (4.46) can be written as

$$
\begin{aligned}
\dot{\alpha}_2^i &= a_1^i \chi_{2d}^i - (\dot{w}_1^i)^\top z_1^i - (\dot{z}_1^i)^\top w_1^i \\
&\quad - c_1^i(-a_1^i \epsilon_1^i - c_1^i \epsilon_1^i + \zeta_2^i). \quad (4.47)
\end{aligned}
$$

The time derivative of (4.43) results in

$$
\dot{\zeta}_2^i = \dot{\epsilon}_2^i - \dot{\alpha}_2^i. \quad (4.48)
$$

From (4.36) and (4.47), (4.48) is written as

$$
\begin{aligned}
\dot{\zeta}_2^i &= -a_2^i \epsilon_2^i - a_2^i \chi_{2d}^i + (w_2^i)^\top z_2^i + u^i - \dot{\chi}_{2d}^i - a_1^i \chi_{2d}^i \\
&\quad + (\dot{w}_1^i)^\top z_1^i + (\dot{z}_1^i)^\top w_1^i + c_1^i(-a_1^i \epsilon_1^i - c_1^i \epsilon_1^i + \zeta_2^i). \\
&\quad\quad (4.49)
\end{aligned}
$$

Thus, from (4.45) and (4.49), the system takes the form

$$
\dot{\epsilon}_1^i = -a_1^i \epsilon_1^i - c_1^i \epsilon_1^i + \zeta_2^i, \quad (4.50)
$$

$$
\begin{aligned}
\dot{\zeta}_2^i &= -a_2^i \zeta_2^i - a_2^i \alpha_2^i - a_2^i \chi_{2d}^i + (w_2^i)^\top z_2^i + u^i - \dot{\chi}_{2d}^i \\
&\quad - a_1^i \chi_{2d}^i + (\dot{w}_1^i)^\top z_1^i + (\dot{z}_1^i)^\top w_1^i \\
&\quad + c_1^i(-a_1^i \epsilon_1^i - c_1^i \epsilon_1^i + \zeta_2^i). \quad (4.51)
\end{aligned}
$$

Proposing the augmented Lyapunov-like function

$$
V_2^i(\xi^i, \tilde{w}^i, \epsilon_1^i, \zeta_2^i) = V^i(\xi^i, \tilde{w}^i) + \frac{1}{2}(\epsilon_1^i)^2 + \frac{1}{2}(\zeta_2^i)^2,
$$

its time derivative, along the trajectories (4.50)–(4.51), is then given by

$$
\begin{aligned}
\dot{V}_2^i(\cdot) &= \dot{V}^i(\xi^i, \tilde{w}^i) + \epsilon_1^i(-a_1^i \epsilon_1^i - c_1^i \epsilon_1^i + \zeta_2^i) \\
&\quad + \zeta_2^i\Big(-a_2^i \zeta_2^i - a_2^i \alpha_2^i - a_2^i \chi_{2d}^i + (w_2^i)^\top z_2^i \\
&\quad + u^i - \dot{\chi}_{2d}^i - a_1^i \chi_{2d}^i + (\dot{w}_1^i)^\top z_1^i + (\dot{z}_1^i)^\top w_1^i \\
&\quad + c_1^i(-a_1^i \epsilon_1^i - c_1^i \epsilon_1^i + \zeta_2^i)\Big).
\end{aligned}
$$

Expanding and rearranging terms yields

$$
\begin{aligned}
\dot{V}_2^i(\cdot) &= \dot{V}^i(\xi^i, \tilde{w}^i) - a_1^i(\epsilon_1^i)^2 - c_1^i(\epsilon_1^i)^2 - a_2^i(\zeta_2^i)^2 \\
&\quad + \zeta_2^i\Big(\epsilon_1^i - a_2^i \alpha_2^i - a_2^i \chi_{2d}^i + (w_2^i)^\top z_2^i \\
&\quad + u^i - \dot{\chi}_{2d}^i - a_1^i \chi_{2d}^i + (\dot{w}_1^i)^\top z_1^i + (\dot{z}_1^i)^\top w_1^i \\
&\quad + c_1^i(-a_1^i \epsilon_1^i - c_1^i \epsilon_1^i + \zeta_2^i)\Big).
\end{aligned}
$$

Thus, selecting the control law as

$$
\begin{aligned}
u^i &= -\epsilon_1^i + a_2^i \alpha_2^i + a_2^i \chi_{2d}^i - (w_2^i)^\top z_2^i + \dot{\chi}_{2d} + a_1^i \chi_{2d} \\
&\quad - (\dot{w}_1^i)^\top z_1^i - (\dot{z}_1^i)^\top w_1^i \\
&\quad - c_1^i(-a_1^i \epsilon_1^i - c_1^i \epsilon_1^i + \zeta_2^i) - c_2^i \zeta_2^i \quad (4.52)
\end{aligned}
$$

this yields

$$
\dot{V}_2^i(\cdot) = \dot{V}^i(\xi, \tilde{w}^i) - (a_1^i + c_1^i)(\epsilon_1^i)^2 - (a_2^i + c_2^i)(\zeta_2^i)^2 \le 0.
$$

Hence, the control law (4.52) asymptotically stabilizes the system.

The block diagram for the decentralized RHONN controller is shown in Fig. 4.1. Fig. 4.2 shows the block diagram for the RHONN controller.

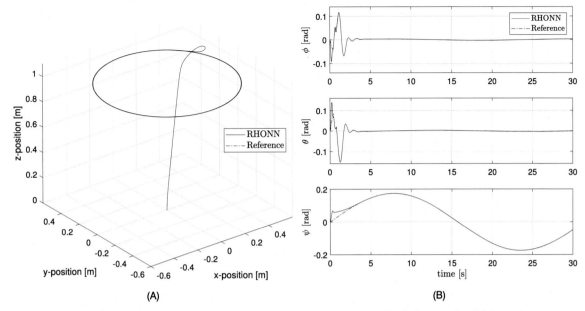

FIG. 4.3 **(A)** Circular trajectory tracking performed by the decentralized RHONN controller. **(B)** Dynamics of the attitude angles.

TABLE 4.1
MSEs from the identification of the quadrotor's dynamics during the performance of circular trajectory tracking.

			Subsystem		
x	y	z	θ	ϕ	ψ
0.003908	0.003906	0.008341	0.003905	0.003892	0.008178

TABLE 4.2
MSEs from the circular trajectory tracking.

			Subsystem		
x	y	z	θ	ϕ	ψ
0.068455	0.022092	0.111574	0.006181	0.004629	0.007953

4.4 RESULTS

In this work, the parameters of the quadrotor are given as $J_x = J_y = 0.03$ kg·m^2, $J_z = 0.04$ kg·m^2, $l = 0.2$ m, $m_q = 1.79$ kg [36].

The performance of the decentralized RHONN control scheme is evaluated through numerical simulation. Fig. 4.3 shows the trajectory tracking task performed by the quadrotor UAV under the decentralized RHONN control scheme. The reference trajectory is defined by $\chi_{1d}^x = 0.5\cos(0.251t)$ and $\chi_{1d}^y = 0.5\sin(0.251t)$.

Table 4.1 exhibits the mean squared errors (MSEs) from the online identification of the quadrotor's dynamics during the performance of the circular trajectory tracking task. MSEs from the performance of the decentralized RHONN controller for trajectory tracking are shown in Table 4.2. Figs. 4.4–4.9 show the identification errors during the performance of the circular trajectory tracking task by the decentralized RHONN controller. Figs. 4.10–4.15 show the respective tracking errors and control signals when performing the circular

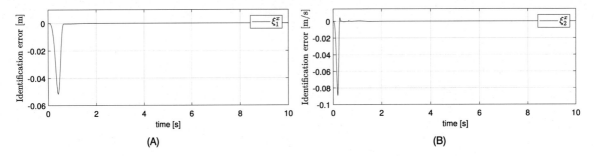

FIG. 4.4 Identification errors of the dynamics from the x-coordinate's subsystem.

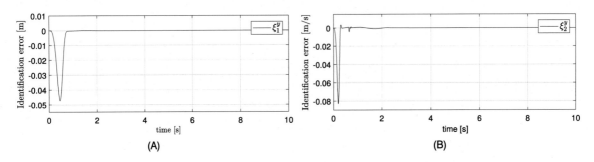

FIG. 4.5 Identification errors of the dynamics from the y-coordinate's subsystem.

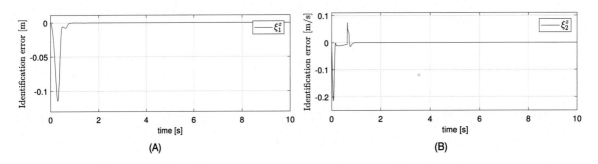

FIG. 4.6 Identification errors of the dynamics from the z-coordinate's subsystem.

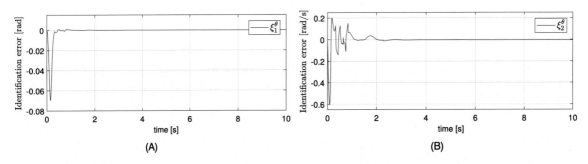

FIG. 4.7 Identification errors of the dynamics from the pitch subsystem.

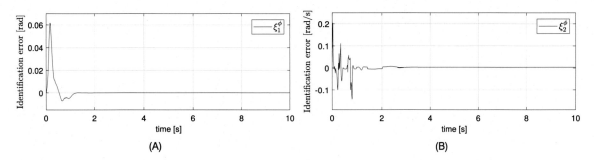

FIG. 4.8 Identification errors of the dynamics from the roll subsystem.

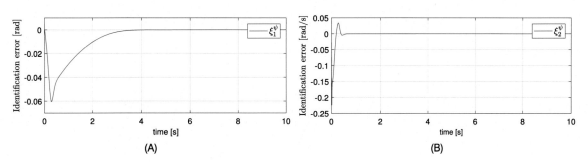

FIG. 4.9 Identification errors of the dynamics from the yaw subsystem.

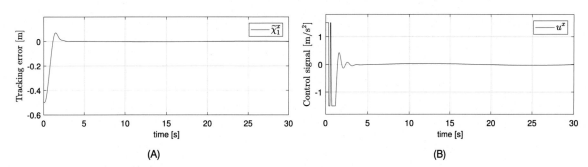

FIG. 4.10 **(A)** Trajectory tracking error for the translational movement on the x-coordinate. **(B)** Decentralized RHONN controller signal.

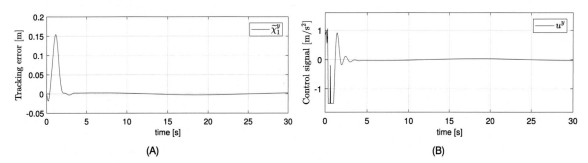

FIG. 4.11 **(A)** Trajectory tracking error for the translational movement on the y-coordinate. **(B)** Decentralized RHONN controller signal.

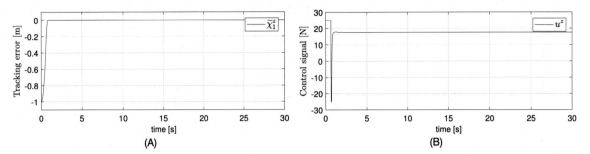

FIG. 4.12 **(A)** Tracking error signal for the translational movement on the z-coordinate. **(B)** Control signal for the altitude subsystem.

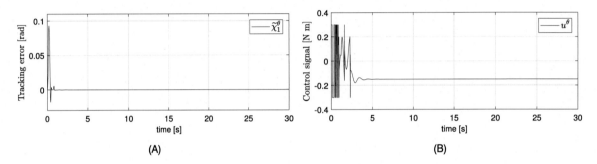

FIG. 4.13 **(A)** Tracking error for the pitch movement. **(B)** Decentralized RHONN controller signal.

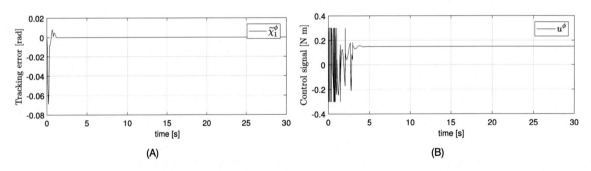

FIG. 4.14 **(A)** Tracking error signal for the roll movement. **(B)** Control signal for the roll subsystem.

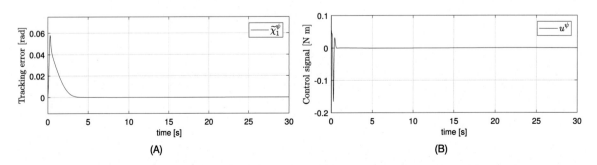

FIG. 4.15 **(A)** Tracking error for the yaw movement. **(B)** Control signal for the yaw subsystem.

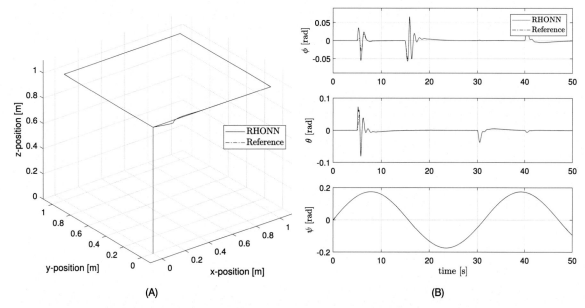

FIG. 4.16 (A) Square-shape trajectory tracking performed by the decentralized RHONN controller.
(B) Dynamics of the attitude angles.

TABLE 4.3
MSEs from the identification of the quadrotor's dynamics during the performance of square-shape trajectory tracking.

Subsystem					
x	y	z	θ	ϕ	ψ
0.003025	0.003021	0.001001	0.003015	0.003012	0.000101

TABLE 4.4
MSEs from the square-shape trajectory tracking.

Subsystem					
x	y	z	θ	ϕ	ψ
0.005785	0.004694	0.086096	0.003100	0.003013	0.000223

trajectory tracking task by the decentralized RHONN controller. Fig. 4.16 shows the tracking task performed by the quadrotor UAV but for a square-shape trajectory. For this latter task, a second-order low-pass filter, with a damping ratio of 0.9 and a natural frequency of 0.55, is used to the reference trajectories χ_{1d}^x and χ_{1d}^y in order to minimize the effect of its derivatives. Table 4.3 exhibits the MSEs from the online identification of the quadrotor's dynamics during the performance of the square-shape trajectory tracking task. Table 4.4 shows

the respective MSEs from performing the square-shape trajectory tracking.

4.5 CONCLUSION

A decentralized RHONN structure is proposed here to identify online, in a series-parallel configuration and using the filtered error learning law, the dynamics from the 6-DOF of a quadrotor UAV. Based on the RHONN subsystems, given in the form of a strict-feedback system, a local neural controller is derived via the backstepping approach. The performance of the overall neural identification and control scheme is validated via simulation results. From the simulation results presented here, it is shown that our proposal performs well. It can be seen that identification errors of the dynamics from the translational and rotational subsystems tend to zero. Due to this, the trajectory tracking task can be carried out by the decentralized RHONN controller. As can be viewed, trajectory tracking errors from each decentralized RHONN controller tend to zero and their respective control signals tend to constant values or to zero after the identification process is finished. Hence, the simulation results validate the viability of the proposed continuous-time decentralized RHONN control scheme.

REFERENCES

1. A.Y. Alanís, E.N. Sánchez, A.G. Loukianov, Real-time output tracking for induction motors by recurrent high-order neural networks control, in: Control and Automation, 2009 17th Mediterranean Conference on, IEEE, 2009, pp. 868–873.
2. A.Y. Alanis, E.N. Sanchez, A.G. Loukianov, E.A. Hernandez, Discrete-time recurrent high order neural networks for nonlinear identification, Journal of the Franklin Institute 347 (7) (2010) 1253–1265.
3. S. Alvarez, C.E. Castañeda, F. Jurado, Neural identification and control using high-order sliding modes, in: Control Automático CLCA 2012, XV Congreso Latinoamericano de, Pontificia Universidad Católica del Perú, 2012, pp. 1–6.
4. M.E. Antonio-Toledo, E.N. Sanchez, A.Y. Alanis, Robust neural decentralized control for a quadrotor UAV, in: Neural Networks (IJCNN), 2016 International Joint Conference on, IEEE, 2016, pp. 714–719.
5. N.A. Bakshi, R. Ramachandran, Indirect model reference adaptive control of quadrotor UAVs using neural networks, in: Intelligent Systems and Control (ISCO), 2016 10th International Conference on, IEEE, 2016, pp. 1–6.
6. S. Bouabdallah, P. Murrieri, R. Siegwart, Design and control of an indoor micro quadrotor, in: Robotics & Automation, 2004 IEEE International Conference on, IEEE, 2004, pp. 4393–4398.
7. O. Bouhali, H. Boudjedir, Neural network control with neuro-sliding mode observer applied to quadrotor helicopter, in: Innovations in Intelligent Systems and Applications (INISTA), 2011 International Symposium on, IEEE, 2011, pp. 24–28.
8. C.E. Castañeda, P. Esquivel, Decentralized neural identifier and control for nonlinear systems based on extended Kalman filter, Neural Networks 31 (2012) 81–87.
9. C.E. Castañeda, A.G. Loukianov, E.N. Sanchez, B. Castillo-Toledo, Discrete-time neural sliding-mode block control for a DC motor with controlled flux, IEEE Transactions on Industrial Electronics 59 (2) (2012) 1194–1207.
10. C.E. Castañeda, D. Mancilla, J.H. García, R.J. Reátegui, G. Huerta, R.C. Zárate, Position control of DC motor based on recurrent high-order neural networks, in: Intelligent Control, 2010 IEEE International Symposium on, IEEE, 2010, pp. 1515–1520.
11. C.E. Castañeda, E.N. Sánchez, A.G. Loukianov, B. Castillo-Toledo, Real-time torque control for a DC motor using recurrent high-order neural networks, in: Intelligent Control, 2009 IEEE International Symposium on, IEEE, 2009, pp. 1809–1814.
12. C.S. Chen, Dynamic structure neural-fuzzy networks for robust adaptive control of robot manipulators, IEEE Transactions on Industrial Electronics 55 (9) (2008) 3402–3414.
13. W. Chen, J. Li, Decentralized output-feedback neural control for systems with unknown interconnections, IEEE Transactions on Systems, Man, and Cybernetics, Part B (2008) 258–266.
14. T. Dierks, S. Jagannathan, Neural network output feedback control of a quadrotor UAV, in: Decision and Control, 2008 47th IEEE Conference on, IEEE, 2008, pp. 3633–3639.
15. T. Dierks, S. Jagannathan, Output feedback control of a quadrotor UAV using neural networks, IEEE Transactions on Neural Networks 21 (1) (2009) 50–66.
16. B. Etkin, L.D. Reid, Dynamics of Flight, Stability and Control, John Wiley & Sons, 1996.
17. M.T. Frye, R.S. Provence, Direct inverse control using an artificial neural network for the autonomous hover of a helicopter, in: Systems, Man and Cybernetics (SMC), 2014 IEEE International Conference on, IEEE, 2014, pp. 4121–4122.
18. R. García-Hernández, J.A. Ruz Hernández, E.N. Sánchez, M. Saad, Real-time decentralized neural control for a five DOF redundant robot, Journal Intelligent Automation and Soft Computing 19 (1) (2013) 23–37.
19. R. García-Hernández, E.N. Sánchez, E. Bayro-Corrochano, M.A. Llama, J.A. Ruz-Hernández, Real-time decentralized neural backstepping control: application to a two DOF robot manipulator, International Journal of Innovative Computing, Information and Control 7 (2) (2011) 965–976.
20. R. García-Hernández, E.N. Sánchez, E. Bayro-Corrochano, V. Santibáñez, J.A. Ruz-Hernández, Real-time decentralized neural block control: application to a two DOF robot manipulator, International Journal of Innovative Computing, Information and Control 7 (3) (2011) 1075–1085.
21. M. Hernandez-Gonzalez, A.Y. Alanis, E.A. Hernandez-Vargas, Decentralized discrete time neural control for a quanser 2-DOF helicopter, Applied Soft Computing 12 (8) (2012) 2462–2469.
22. S. Huang, K.K. Tan, T.H. Lee, Decentralized control design for large-scale systems with strong interconnections using neural networks, IEEE Transactions on Automatic Control 48 (5) (2003) 805–810.
23. P.A. Ioannou, A. Datta, Robust adaptive control: a unified approach, Proceedings of the IEEE 79 (12) (1991) 1736–1768.

24. Y. Jin, Decentralized adaptive fuzzy control of robot manipulator, IEEE Transactions on Systems, Man, and Cybernetics, Part B (Cybernetics) 28 (1) (1998) 47–57.
25. F. Jurado, M.A. Flores, C.E. Castañeda, Continuous-time neural control for a 2 DOF vertical robot manipulator, in: Electrical Engineering, Computing Science and Automatic Control (CCE), 2011 8th International Conference on, IEEE, 2011, pp. 1–6.
26. F. Jurado, M.A. Flores, V. Santibáñez, M.A. Llama, C.E. Castañeda, Continuous-time neural identification for a 2 DOF vertical robot manipulator, in: Electronics, Robotics and Automotive Mechanics Conference, 2011 IEEE, IEEE, 2011, pp. 77–82.
27. F. Jurado, S. Lopez, A wavelet neural control scheme for a quadrotor unmanned aerial vehicle, Philosophical Transactions of The Royal Society A 376 (2018) 20170248, https://doi.org/10.1098/rsta.2017.0248.
28. E.B. Kosmatopoulos, M.M. Polycarpou, M.A. Christodoulou, P.A. Ioannou, High-order neural network structures for identification of dynamical systems, IEEE Transactions on Neural Networks 6 (2) (1995) 422–431.
29. E. Lavretsky, K.A. Wise, Robust and Adaptive Control with Aerospace Applications, Springer-Verlag, 2013.
30. M. Liu, Decentralized control of robot manipulators: nonlinear and adaptive approaches, IEEE Transactions on Automatic Control 44 (2) (1999) 357–363.
31. T. Madani, A. Benallegue, Adaptive control via backstepping technique and neural networks of a quadrotor helicopter, in: 17th IFAC World Congress (IFAC'08), 2008.
32. H. Neijmeijer, A. van der Schaft, Nonlinear Dynamical Control Systems, Springer-Verlag, 1990.
33. M.L. Ni, M.J. Er, Decentralized control of robot manipulators with coupling and uncertainties, in: Proceedings of the 2000 American Control Conference ACC, vol. 5, 2000, pp. 3326–3330.
34. C. Nicol, C.J.B. Macnab, A. Ramirez-Serrano, Robust neural network control of a quadrotor helicopter, in: Electrical and Computer Engineering, 2008 Canadian Conference on, IEEE, 2008, pp. 1233–1238.
35. F. Ornelas-Tellez, A.G. Loukianov, E.N. Sanchez, E.J. Bayro-Corrochano, Decentralized neural identification and control for uncertain nonlinear systems: application to planar robot, Journal of the Franklin Institute 347 (6) (2010) 1015–1034.
36. Quanser Inc., User Manual QBall 2 for QUARC, 2014.
37. J.D. Rios, A.Y. Alanis, J. Rivera, M. Hernandez-Gonzalez, Real-time discrete neural identifier for a linear induction motor using a dspace ds1104 board, in: Neural Networks (IJCNN), Proceedings of International Joint Conference on, IEEE, 2013, pp. 1–6.
38. G.A. Rovithakis, M.A. Christodoulou, Adaptive Control with Recurrent High-Order Neural Networks, Springer-Verlag, 2000.
39. R. Safaric, J. Rodic, Decentralized neural-network sliding-mode robot controller, in: IEEE Industrial Electronics Society IECON, 2000 26th Annual Conference of the, vol. 2, IEEE, 2000, pp. 906–911.
40. E.N. Sanchez, A. Gaytan, M. Saad, Decentralized neural identification and control for robotics manipulators, in: Intelligent Control, 2006 IEEE International Symposium on, IEEE, 2006, pp. 1614–1619.
41. B.L. Stevens, F.L. Lewis, E.N. Johnson, Aircraft Control and Simulation, John Wiley & Sons, 2016.
42. L.A. Vázquez, F. Jurado, C.E. Castañeda, V. Santibáñez, Real-time decentralized neural control via backstepping for a robotic arm powered by industrial servomotors, IEEE Transactions on Neural Networks and Learning Systems 29 (2) (2018) 419–426.
43. R. Vepa, Nonlinear Control of Robots and Unmanned Aerial Vehicles: An Integrated Approach, CRC Press, 2017.
44. H. Voos, Nonlinear and neural network-based control of a small four-rotor aerial robot, in: Advanced Intelligent Mechatronics, International Conference on, IEEE/ASME, 2007.
45. Y. Wang, H. Zhang, D. Han, Neural network adaptive inverse model control method for quadrotor UAV, in: 2016 35th Chinese Control Conference (CCC), 2016.

CHAPTER 5

Adaptive PID Controller Using a Multilayer Perceptron Trained With the Extended Kalman Filter for an Unmanned Aerial Vehicle

JAVIER GOMEZ-AVILA, PHD

5.1 INTRODUCTION

The objective of control is to take a dynamical system from an initial state to a final state in a finite amount of time. Despite some classical approaches such as proportional integral derivative (PID) controllers are widely used in the industry, where systems are assumed to be linear and time-invariant, they are not suitable for highly nonlinear systems where uncertainties and unknown time delays are present. In practice, systems are nonlinear, present unmodeled dynamics, and, because of uncertainties (commonly due to the operation itself such as changes in rotor efficiency), it is impossible to control with classical control approaches.

On the other hand, approaches like backstepping control, designed as a powerful tool to control nonlinear systems [2,9,19], require analytic calculation of the partial derivatives of the stabilizing functions, which becomes impractical as the order of the system grows [4]. Conversely, some artificial intelligence (AI) paradigms, like artificial neural networks (ANNs), have been used to control nonlinear systems. One of the advantages of using ANNs is that the controller will have the adaptability and learning capabilities of the neural network [5], and as a result, the system will be able to adapt to unmodeled dynamics, communication time delay, parametric uncertainties, external disturbances, and actuator saturation, among others [12]. In this chapter, a multilayer perceptron (MLP) is proposed to solve this problem and control an unmanned aerial vehicle (UAV) trained online. With this in mind, it is necessary to implement a training algorithm that does not require a complete data set of system outputs since it will be generated during experimentation.

The Kalman filter is an optimal estimator which infers parameters based on noisy and inaccurate observations. Its solution is recursive and as a consequence, the filter can process data as soon as they arrive and predict the next value without the need of having the complete data set of observations [8], making it faster and convenient for online applications in contrast with other methods such as batch processing.

Some training algorithms for ANNs, like gradient descent, recursive least squares, and backpropagation, are particular cases of the Kalman filter, and feasible results have been obtained [10,14,18]. In fact, the extended Kalman filter (EKF) has been successfully applied to the estimation of ANN parameters [3,16,17].

5.2 KALMAN FILTER

The Kalman filter [8] is one of the most popular data fusion algorithms, which predict a future state vector based on prior knowledge of the system and the current measurement of the state (Fig. 5.1). It has been regarded as the optimal solution to data prediction problems.

From [1], the filter assumes that the state of a system will evolve according to

$$w(k+1) = F_{k+1,k} w(k) + u_t, \qquad (5.1)$$

where $F_{k+1,k} \in \Re^{L \times L}$ is the state transition matrix from iteration k to $k+1$; $w(k) \in \Re^L$ is the state vector and $u(k) \in \Re^L$ represents the process noise vector assumed to be zero mean Gaussian with covariance matrix defined as

$$E\left\{u(n) u^T(k)\right\} = \begin{cases} Q(k) & \text{if } n = k, \\ 0 & \text{if } n \neq k. \end{cases} \qquad (5.2)$$

The measurements of the system can be performed according to

$$y(k) = H(k) w(k) + v(k), \qquad (5.3)$$

Artificial Neural Networks for Engineering Applications. https://doi.org/10.1016/B978-0-12-818247-5.00014-9

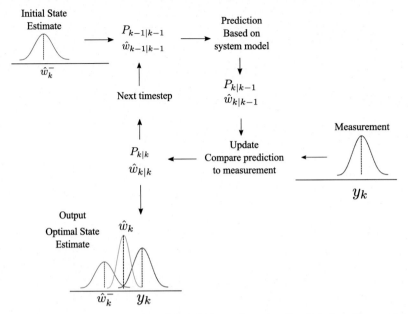

FIG. 5.1 Kalman filter outline. A prediction of the initial state based on the state transition model is compared with the measurement.

where $y \in \Re^m$ is the vector measurements, $H \in \Re^{m \times L}$ is the transformation matrix that maps the state vector into the measurement domain, and $v \in \Re^m$ contains the measurement noise which is assumed to be zero mean Gaussian white noise with covariance R_t defined as

$$E\left\{v(n) v^T(k)\right\} = \begin{cases} R(k) & \text{if } n = k, \\ 0 & \text{if } n \neq k. \end{cases} \quad (5.4)$$

The state vector $w(k)$ can be predicted changing between two stages: prediction and update. The prediction state equations include the state vector (5.5) and covariance matrix (5.6) propagation defined as

$$\hat{w}^-(k) = F_{k+1,k}\,\hat{w}^-(k-1), \quad (5.5)$$

$$P^-(k) = F_{k+1,k}\,P(k-1)\,F^T(k, k-1) + Q(k-1). \quad (5.6)$$

The update stage consists of the computation of the Kalman gain K and the update of the state vector $w(k)$ and the covariance matrix $P(k)$. We have

$$K(k) = P^-(k)\,H^T(k)\left[H(k)\,P^-(k)\,H^T(k) + R(k)\right]^{-1}, \quad (5.7)$$

$$\hat{w}(k) = \hat{w}^-(k) + K(k)\left(y(k) - H(k)\,\hat{w}^-(k)\right), \quad (5.8)$$

$$P(k) = (I - K(k)\,H(k))\,P^-(k). \quad (5.9)$$

If $w(0)$, $u(k)$, and $v(k)$ are normal random variables, then the Kalman filter estimation is optimal for $w(k)$ [11]. It is also important to initialize $P(0) \neq 0$, otherwise the initialization of the states would be assumed perfect and the Kalman gain K in (5.7) will not be updated. Nevertheless, the Kalman filter assumes a linear model while most of the systems (including the ANN) are normally nonlinear. In order to estimate the states of a nonlinear model, it is necessary to use an alternative algorithm which implies a linearization method known as the extended Kalman filter (EKF).

5.2.1 Extended Kalman Filter

Consider a nonlinear dynamic system described by

$$w(k+1) = f(k, w(k)) + u(k), \quad (5.10)$$

$$y(k) = h(k, w(k)) + v(k), \quad (5.11)$$

where, additionally to Kalman filter equations, $f(k, w(k))$ represents the nonlinear (and possibly time-variant) transition function and $h(k, w(k))$ is the observation nonlinear vector function. These functions $f(\cdot)$ and $h(\cdot)$ cannot be directly applied to the covariance; instead, it is necessary to linearize the system described by (5.10) and (5.11) and then apply Eqs. (5.5)–(5.9) [7].

The linearization is carried out, firstly, computing

$$F_{k+1,k} = \left. \frac{\partial f(k, w(k))}{\partial w} \right|_{w=\hat{w}(k)}, \quad (5.12)$$

$$H_{k+1,k} = \left. \frac{\partial h(k, w(k))}{\partial w} \right|_{w=\hat{w}^-(k)}. \quad (5.13)$$

Once F and H have been evaluated, functions $f(k, w(k))$ and $h(k, w(k))$ are approximated using Taylor series as follows:

$$f(k, w(k)) \approx F(w, \hat{w}(k)) + F_{k+1,k}(w, \hat{w}(k)), \quad (5.14)$$

$$h(k, w(k)) \approx H(w, \hat{w}^-(k)) + H_{k+1,k}(w, \hat{w}^-(k)). \quad (5.15)$$

Using (5.14) and (5.15) to approximate the state equations (5.10) and (5.11) we have

$$w(k+1) \approx F_{k+1,k} w(k) + u(k) + d(k), \quad (5.16)$$

$$\bar{y}(k) = H(k) w(k) + v(k), \quad (5.17)$$

where the new variables $d(k)$ and $\bar{y}$ are defined as

$$d(k) = f(w, \hat{w}(k)) - F_{k+1,k} \hat{w}(k), \quad (5.18)$$

$$\bar{y}(k) = y(k) - \left\{ h(w, \hat{w}^-(k)) - H(k) \hat{w}^-(k) \right\}. \quad (5.19)$$

Then the state vector $w(k)$ of the linear system (5.10) and (5.11) is defined by the following set of equations.
Prediction:

$$\hat{w}^-(k) = f(k, \hat{w}(k-1)), \quad (5.20)$$

$$P^-(k) = F_{k,k-1} P(k-1) F_{k,k-1}^T + Q(k-1). \quad (5.21)$$

Update:

$$K(k) = P^-(k) H^T(k) \left[H(k) P^-(k) H^T(k) + R(k) \right]^{-1}, \quad (5.22)$$

$$\hat{w}(k) = \hat{w}^-(k) + K(k) y(k) - h(k, \hat{w}^-(k)), \quad (5.23)$$

$$P(k) = (I - K(k) H(k)) P^-(k). \quad (5.24)$$

5.3 MLP TRAINED WITH THE EKF

The training process of the MLP can be seen as an estimation problem for a nonlinear system and, as a consequence, it can be solved with the EKF [6,13,18]. When using the EKF, the network weights are the states that the filter estimates and the output of the MLP is the measurement used by the EKF. The objective is to find the optimal weights that minimize the prediction error [1].

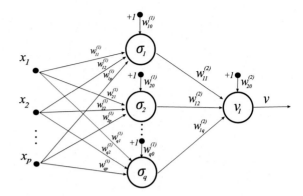

FIG. 5.2 Multilayer perceptron architecture. The network has p inputs and q nodes in the hidden layer. The weights from the input layer to the hidden layer are denoted by $w_{ij}^{(1)}$ while the weights from the hidden layer to the output layer are described by $w_{ij}^{(2)}$.

Consider an MLP with L weights and m output nodes. The neural network can be modeled with

$$w(k+1) = w(k), \quad (5.25)$$

$$\hat{y}(k) = h(w(k), u(k)), \quad (5.26)$$

where $w(k)$ is the state vector, $u(k)$ is the input vector, $\hat{y}$ is the output vector, and h is the nonlinear output function. Using the Kalman filter equations, we have

$$K(k) = P(k) H^T(k) \left[R(k) + H(k) P(k) H^T(k) \right]^{-1}, \quad (5.27)$$

$$w(k+1) = w(k) + \eta K(k) \left[y(k) - \hat{y}(k) \right], \quad (5.28)$$

$$P(k) = P(k) - K(k) H(k) P(k) + Q(k), \quad (5.29)$$

where η is the learning rate, L is the total number of weights, $P(k) \in \Re^{L \times L}$ and $P(k+1) \in \Re^{L \times L}$ are the prediction error covariance matrices in k and $k+1$, $K(k) \in \Re^{L \times m}$ is the Kalman gain matrix, $y \in \Re^m$ is the system output, $\hat{y}$ is the network output, $Q \in \Re^{L \times L}$ is the process noise covariance matrix, $R \in \Re^{m \times m}$ represents the measurement covariance error, $w \in \Re^L$ is the weights (state) vector, and $H \in \Re^{m \times L}$ contains the partial derivatives of each output of the neural network $\hat{y}$ with respect to each weight w_j; it is defined as

$$H_{ij}(k) = \left[\frac{\partial \hat{y}_i(k)}{\partial w_j(k)} \right]_{w(k)=\hat{w}(k+1)}, i = 1 \ldots m, j = 1 \ldots L. \quad (5.30)$$

Consider the MLP in Fig. 5.2 with one hidden layer and one node at the output layer, where p denotes the

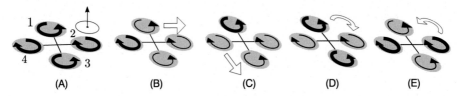

FIG. 5.3 Quadrotor motion. The arrow width in the propeller is proportional to its rotation. White arrow describes the motion of the quadrotor.

number of inputs to the network and q the number of nodes in the hidden layer. The output of the network is defined by

$$\sigma_i = \frac{1}{1+e^{-n_i}} \quad i = 1 \ldots q, \qquad (5.31)$$

$$n_i = \sum_{j=0}^{p} w_{ij}^{(1)} x_j \quad x_0 = +1, \qquad (5.32)$$

$$v_1 = \sum_{k=0}^{q} w_{1k}^{(2)} u_k \quad u_0 = +1, \qquad (5.33)$$

$$\hat{y} = v_1. \qquad (5.34)$$

Finally, vector H can be expressed as

$$H = \frac{\partial \hat{y}}{\partial w} = \left[\begin{array}{cccc} \frac{\partial \hat{y}}{\partial w_{10}^{(1)}} & \frac{\partial \hat{y}}{\partial w_{11}^{(1)}} & \cdots & \frac{\partial \hat{y}}{\partial w_{1q}^{(2)}} \end{array} \right], \qquad (5.35)$$

where

$$
\begin{aligned}
&\frac{\partial \hat{y}}{\partial w_{10}^{(1)}} = \frac{w_{11}^{(2)} e^{-n_1}}{\left(1-e^{-n_1}\right)^2} x_0, \ \frac{\partial \hat{y}}{\partial w_{11}^{(1)}} = \frac{w_{11}^{(2)} e^{-n_1}}{\left(1-e^{-n_1}\right)^2} x_1, \ \cdots \ \frac{\partial \hat{y}}{\partial w_{1p}^{(1)}} = \frac{w_{11}^{(2)} e^{-n_1}}{\left(1-e^{-n_1}\right)^2} x_p, \\
&\frac{\partial \hat{y}}{\partial w_{20}^{(1)}} = \frac{w_{12}^{(2)} e^{-n_2}}{\left(1-e^{-n_2}\right)^2} x_0, \ \frac{\partial \hat{y}}{\partial w_{21}^{(1)}} = \frac{w_{12}^{(2)} e^{-n_2}}{\left(1-e^{-n_2}\right)^2} x_1, \ \cdots \ \frac{\partial \hat{y}}{\partial w_{2p}^{(1)}} = \frac{w_{12}^{(2)} e^{-n_2}}{\left(1-e^{-n_2}\right)^2} x_p, \\
&\quad \vdots \qquad\qquad\qquad \vdots \qquad\quad \vdots \qquad\qquad \vdots \\
&\frac{\partial \hat{y}}{\partial w_{q0}^{(1)}} = \frac{w_{1q}^{(2)} e^{-n_q}}{\left(1-e^{-n_q}\right)^2} x_0, \ \frac{\partial \hat{y}}{\partial w_{q1}^{(1)}} = \frac{w_{1q}^{(2)} e^{-n_q}}{\left(1-e^{-n_q}\right)^2} x_1, \ \cdots \ \frac{\partial \hat{y}}{\partial w_{qp}^{(1)}} = \frac{w_{1q}^{(2)} e^{-n_q}}{\left(1-e^{-n_q}\right)^2} x_p, \\
&\frac{\partial \hat{y}}{\partial w_{10}^{(2)}} = 1, \qquad \frac{\partial \hat{y}}{\partial w_{11}^{(2)}} = \frac{1}{1+e^{-n_1}}, \quad \cdots \quad \frac{\partial \hat{y}}{\partial w_{1q}^{(2)}} = \frac{1}{1+e^{-n_q}}.
\end{aligned}
$$
$$ (5.36) $$

If we define a new function $\gamma(\cdot)$ as

$$\gamma(n_i) = \frac{w_{1i}^{(2)} e^{-n_i}}{\left(1+e^{-n_i}\right)^2}, \quad i = 1 \ldots q, \qquad (5.37)$$

then the vector H for the MLP shown in Fig. 5.2 with sigmoid activation functions for the hidden layers and linear function for the output node can be expressed as

$$H = \left[\begin{array}{ccccc} \gamma(n_1) x_0 & \cdots & \gamma(n_1) x_p & \gamma(n_2) x_0 & \cdots \\ & \gamma(n_q) x_p & u_0 & u_1 & \cdots & u_q \end{array} \right]. \qquad (5.38)$$

5.4 UAV CONTROLLED WITH AN MLP

The control of UAVs is an important task in mobile robotics because of their extensive use and the advantages over other kind of mobile robots like ground vehicles. In contrast to ground vehicles, flying robots have gained great interest in recent decades because of their ability to move in three-dimensional space and travel longer distances in less time. UAVs have important applications for both military and civil markets such as rescue operations and surveillance. Recent technological advances in lightweight sensors, actuators, and embedded electronic systems have allowed the development of vehicles based on vertical take-off and landing (VTOL). The quadrotor is an aerial vehicle based on VTOL and is one of the most popular platforms due to its maneuverability, stationary flight, low-speed fly, and both indoor and outdoor usage. However, quadrotors are highly nonlinear underactuated systems and, therefore, difficult to control with conventional methods.

5.4.1 Quadrotor Dynamic Modeling

Consider a vehicle with two pairs of propellers (1,3) and (2,4) in cross-configuration and turning in opposite directions, as shown in Fig. 5.3. Increasing or decreasing the speed of the four rotors results in vertical motion of the vehicle (Fig. 5.3A). Changing the speed of the rotors of the pair (2,4) generates a roll rotation (Fig. 5.3B). On the other hand, changing the speed of the rotors of the pair (1,3) will produce a pitch rotation (Fig. 5.3C). Roll and pitch rotations will be traduced to a lateral motion. Yaw rotation is generated when increasing or decreasing one pair and it is the result of the difference of the countertorque between the pairs (Fig. 5.3D and Fig. 5.3E).

Without loss of generality, let us consider that the center of mass and the body fixed frame origin coincide, as in Fig. 5.4. The quadrotor orientation in space is given by a rotation matrix $R \in SO(3)$ from the body fixed frame to the inertial frame. The dynamics of a rigid

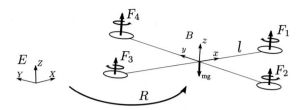

FIG. 5.4 Quadrotor configuration. B represents the quadrotor fixed frame and E the inertial frame.

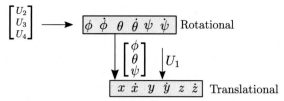

FIG. 5.5 The terms U_2, U_3, and U_4 are inputs for the rotational subsystem; U_1, roll, pitch, and yaw are inputs for the following translation subsystem.

body under external forces is expressed as

$$
\begin{bmatrix} mI_{3\times3} & 0 \\ 0 & I \end{bmatrix} \begin{bmatrix} \dot{V} \\ \dot{\omega} \end{bmatrix} + \begin{bmatrix} \omega \times mV \\ \omega \times I\omega \end{bmatrix} = \begin{bmatrix} F \\ \tau \end{bmatrix},
$$
(5.39)

where I is the inertia matrix, V the body linear speed vector, and ω the angular speed.

The quadrotor equations of motion can be expressed as

$$
\begin{aligned}
\dot{\zeta} &= v, \\
\dot{v} &= -ge_3 + R_{e3}\left(\tfrac{b}{m}\sum\Omega_i^2\right), \\
\dot{R} &= R\hat{\omega}, \\
I\dot{\omega} &= -\omega \times I\omega - \sum J_r\left(\omega \times e_3\right)\Omega_i + \tau_a,
\end{aligned}
$$
(5.40)

where ζ is the position vector, R the rotation matrix, $\hat{\omega}$ the skew symmetric matrix, Ω_i the rotor speed, I the body inertia, J_r the rotor inertia, b the thrust factor, d the drag factor, l the distance from the body fixed frame origin to the rotor, and τ_a the torque applied to the quadrotor; this is expressed as

$$
\tau_a = \begin{pmatrix} lb\left(\Omega_4^2 - \Omega_2^2\right) \\ lb\left(\Omega_3^2 - \Omega_1^2\right) \\ d\left(\Omega_2^2 + \Omega_4^2 - \Omega_1^2 - \Omega_3^2\right) \end{pmatrix}.
$$
(5.41)

The full quadrotor dynamic model is defined as

$$
\begin{aligned}
\ddot{x} &= (\cos(\phi)\sin(\theta)\sin(\psi) + \sin(\phi)\sin(\psi))\tfrac{U_1}{m}, \\
\ddot{y} &= (\cos(\phi)\sin(\theta)\sin(\psi) - \sin(\phi)\sin(\psi))\tfrac{U_1}{m}, \\
\ddot{z} &= -g + (\cos(\phi)\cos(\theta))\tfrac{U_1}{m}, \\
\ddot{\phi} &= \dot{\theta}\dot{\psi}\left(\tfrac{I_y - I_z}{I_x}\right) - \tfrac{J_r}{I_x}\dot{\theta}\Omega + \tfrac{l}{I_x}U_2, \\
\ddot{\theta} &= \dot{\phi}\dot{\psi}\left(\tfrac{I_z - I_x}{I_y}\right) + \tfrac{J_r}{I_y}\dot{\phi}\Omega + \tfrac{l}{I_y}U_3, \\
\ddot{\psi} &= \dot{\phi}\dot{\theta}\left(\tfrac{I_x - I_y}{I_z}\right) + \tfrac{U_4}{I_z},
\end{aligned}
$$
(5.42)

with U_i being the system's inputs and Ω the changing attitude angle, which is part of the gyroscopic effects induced by the propellers. Gyroscopic effects provide a more accurate model; nevertheless, they play insignificant roles in the overall attitude of the quadrotor [15]. The inputs of the system are defined as

$$
\begin{aligned}
U_1 &= b\left(\Omega_1^2 + \Omega_2^2 + \Omega_3^2 + \Omega_4^2\right), \\
U_2 &= b\left(\Omega_4^2 - \Omega_2^2\right), \\
U_3 &= b\left(\Omega_3^2 - \Omega_1^2\right), \\
U_4 &= b\left(\Omega_2^2 + \Omega_4^2 - \Omega_1^2 - \Omega_3^2\right), \\
\Omega &= \Omega_2 + \Omega_4 - \Omega_1 - \Omega_3.
\end{aligned}
$$
(5.43)

The quadrotor is a rotating rigid body with six degrees of freedom and rotational and translational dynamics. Inputs U_i and their relation with both subsystems are shown in Fig. 5.5. The dynamic model is the same when working with different configurations of VTOL vehicles; however, the rotor combinations in (5.41) and (5.43) will change if the configuration is modified (rotors do not coincide with coordinate axis) or more rotors are added (e.g., hexarotor [12]).

5.4.2 Quadrotor Control Scheme

For the next simulations an MLP will be used with an architecture as shown in Fig. 5.2. The network has two inputs: the error between the desired value and the output measured from the system and the derivative of the error in order to have information about the rate of change of the process variable (with the objective of having the same effect as the derivative gain in a conventional PID controller). The network has one hidden layer with 10 nodes and one neuron at the output layer, and there are four MLP modules, one for each controllable degree of freedom (Fig. 5.6). The outputs of the four MLP modules represent the control inputs U_i from (5.43).

For a first experiment, the desired altitude is constant at 4 m while the trajectory in x is a square signal and the

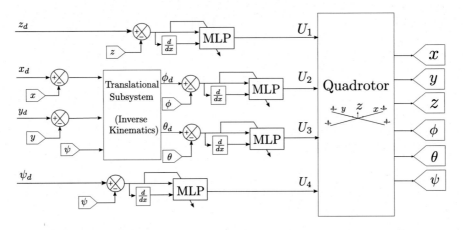

FIG. 5.6 Control of quadrotor with MLP. There is one MLP module for each degree of freedom to control.

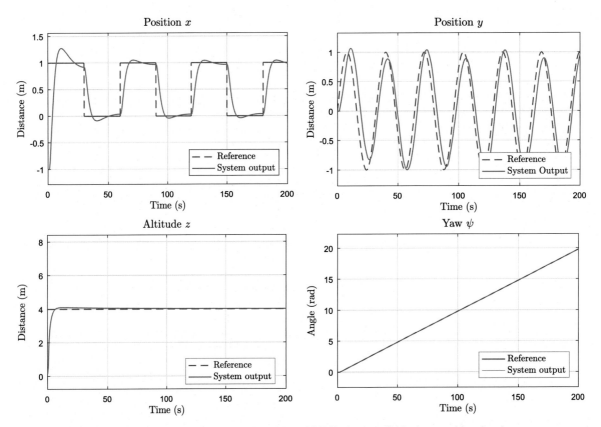

FIG. 5.7 Simulation of a quadrotor controlled with an MLP. Each controllable degree of freedom is controlled with an MLP.

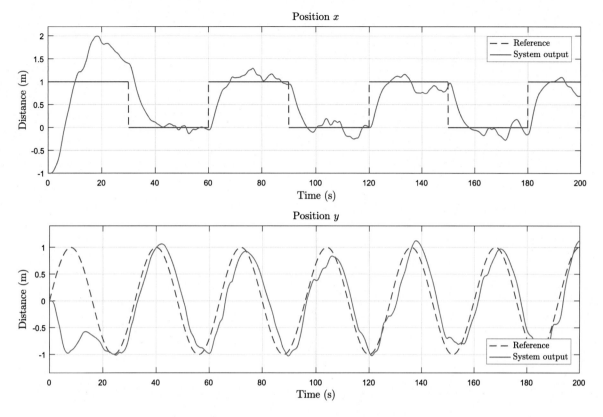

FIG. 5.8 Simulation of a quadrotor controlled with an MLP with disturbances in roll (ϕ) and pitch (θ).

trajectory in y is a sine wave. Desired yaw (ψ) is a ramp with a constant slope value of 0.1, as can be seen in the results shown in Fig. 5.7.

The test of Fig. 5.7 is now repeated but this time simulating disturbance in roll and pitch angles. The results of position in x- and y-directions are reported in Fig. 5.8.

For a third experiment, suppose that the quadrotor grabs something during flight and the mass of the quadrotor is increased 50% at $t = 40$ s. Fig. 5.9 shows the position on the z-axis with different learning rates $\eta = 0.005$, $\eta = 0.05$, and $\eta = 0.5$. As can be seen, the integral effect of a PID controller can be achieved thanks to the adaptability of the neural network.

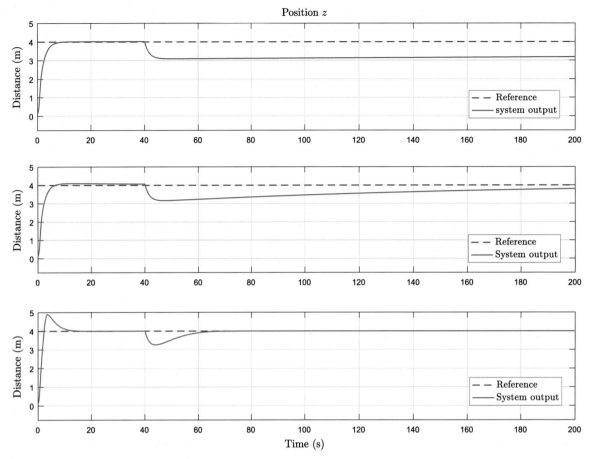

FIG. 5.9 Simulation of a quadrotor changing its mass during flight. The mass increases 50% and the subfigures report the system output with different learning rates: 0.005, 0.05, 0.5, respectively.

REFERENCES

1. E.N.S. Camperos, A.Y.A. García, Redes neuronales: conceptos fundamentales y aplicaciones a control automático, Pearson Educación, 2006.
2. M. Chen, S.S. Ge, B.V.E. How, Robust adaptive neural network control for a class of uncertain mimo nonlinear systems with input nonlinearities, IEEE Transactions on Neural Networks 21 (5) (2010) 796–812.
3. J.F. de Freitas, M. Niranjan, A.H. Gee, Hierarchical Bayesian models for regularization in sequential learning, Neural Computation 12 (4) (2000) 933–953.
4. W. Dong, J.A. Farrell, M.M. Polycarpou, V. Djapic, M. Sharma, Command filtered adaptive backstepping, IEEE Transactions on Control Systems Technology 20 (3) (2012) 566–580.
5. S.S. Ge, J. Zhang, T.H. Lee, Adaptive neural network control for a class of mimo nonlinear systems with disturbances in discrete-time, IEEE Transactions on Systems, Man, and Cybernetics, Part B (Cybernetics) 34 (4) (2004) 1630–1645.
6. S. Haykin, Neural Networks: A Comprehensive Foundation, 1999, Mc Millan, New Jersey, 2010.
7. S.S. Haykin, et al., Kalman Filtering and Neural Networks, Wiley Online Library, 2001.
8. R.E. Kalman, A new approach to linear filtering and prediction problems, Journal of Basic Engineering 82 (1) (1960) 35–45.
9. P.V. Kokotovic, The joy of feedback: nonlinear and adaptive, IEEE Control Systems 12 (3) (1992) 7–17.
10. H. Leung, S. Haykin, The complex backpropagation algorithm, IEEE Transactions on Signal Processing 39 (9) (1991) 2101–2104.
11. F.L. Lewis, F. Lewis, Optimal Estimation: With an Introduction to Stochastic Control Theory, Wiley, New York [etc.], 1986.
12. C. Lopez-Franco, J. Gomez-Avila, A.Y. Alanis, N. Arana-Daniel, C. Villaseñor, Visual servoing for an autonomous hexarotor using a neural network based PID controller, Sensors 17 (8) (2017), http://www.mdpi.com/1424-8220/17/8/1865.

13. G.V. Puskorius, L.A. Feldkamp, Parameter-based Kalman filter training: theory and implementation, in: Kalman Filtering and Neural Networks, 2001, p. 23.

14. D.W. Ruck, S.K. Rogers, M. Kabrisky, P.S. Maybeck, M.E. Oxley, Comparative analysis of backpropagation and the extended Kalman filter for training multilayer perceptrons, IEEE Transactions on Pattern Analysis and Machine Intelligence 14 (6) (1992) 686–691.

15. M.D. Schmidt, Simulation and control of a quadrotor unmanned aerial vehicle, 2011.

16. S. Singhal, L. Wu, Training feed-forward networks with the extended Kalman algorithm, in: Acoustics, Speech, and Signal Processing, 1989, ICASSP-89, 1989 International Conference on, IEEE, 1989, pp. 1187–1190.

17. R.J. Williams, Some Observations on the Use of the Extended Kalman Filter as a Recurrent Network Learning Algorithm, Citeseer, 1992.

18. R.J. Williams, Training recurrent networks using the extended Kalman filter, in: Neural Networks, 1992, IJCNN, International Joint Conference on, vol. 4, IEEE, 1992, pp. 241–246.

19. B. Xu, Robust adaptive neural control of flexible hypersonic flight vehicle with dead-zone input nonlinearity, Nonlinear Dynamics 80 (3) (2015) 1509–1520.

CHAPTER 6

Support Vector Regression for Digital Video Processing

GEHOVÁ LÓPEZ-GONZÁLEZ, PHD

6.1 INTRODUCTION

A video is a sequence of digital images. A digital image is a discrete multidimensional signal, as seen in Fig. 6.1. The representation is a set of matrices, one for each channel. The channels can represent intensity (like grayscale or different colors) and transparency, among other things. In consequence, digital image processing is a subcategory of digital signal processing. One of its areas of application is the improvement of image quality for human perception.

There are no objective criteria for image enhancement. The viewer must evaluate how well a method works for a particular case.

The method proposed in this work uses regression analysis for three tasks. We give an explanation of these tasks in the following subsections. We obtained images with the use of the software Scilab [8].

6.1.1 Adaptive Filter

Filtering is a process of noise removal from a measured process to reveal or enhance information. A linear filter is a normalized square matrix of size n, called kernel. The kernel represents a function, like a box, e.g.,

$$box = \frac{1}{9} \begin{bmatrix} 1 & 1 & 1 \\ 1 & 1 & 1 \\ 1 & 1 & 1 \end{bmatrix}, \tag{6.1}$$

or Gaussian, e.g.,

$$Gauss = \frac{1}{273} \begin{bmatrix} 1 & 4 & 7 & 4 & 1 \\ 4 & 16 & 26 & 16 & 4 \\ 7 & 26 & 41 & 26 & 7 \\ 4 & 16 & 26 & 16 & 4 \\ 1 & 4 & 7 & 4 & 1 \end{bmatrix}. \tag{6.2}$$

The filter modifies the image through convolution. Convolution is the process of adding each element of the image to its local neighbors. The kernel determines the weight of each element. This behavior allows the filter to remove noise, but it also blurs the details of the image. This raises a question. Is it possible to create an intelligent filter that adapts the weights for each pixel?

Adaptive filters are the answer to this question. The idea is varying the values of the kernel as it slides across the image. They are able to adapt themselves to the local characteristics of the image. In essence, an adaptive filter combines a linear filter with another method. It behaves as a self-adjusting digital filter.

The adaptive filter must be capable of recognizing outliers or atypical pixels. There are many approaches to do this. Some use an error-based metric. The minimal mean square error filter (MMSE) [5] computes the error estimate

$$\widehat{f} = E(f|g), \tag{6.3}$$

where f is the original image, g is the degraded observation, and E is the conditional mean estimate.

Others combine linear and nonlinear filters. The alpha-trimmed mean [11] is an hybrid of a media filter and the median filter. It discards a number of α elements in each neighborhood. It then computes the media with the rest. For example, with $\alpha = 2$, we can have

200	137	127
143	141	132
111	120	67

$\rightarrow$ | 67̶ | 111 | 120 | 127 | 132 | 137 | 141 | 143 | 2̶0̶0̶ |

$$111 + 120 + 127 + 132 + 137 + 141 + 143$$
$$= 911 \rightarrow 911/7 = 130.14 \rightarrow 130.$$

Other methods focus on impulse noise [4]. The ranked order–based adaptive median filter (RAMF) detects impulses in the center pixel. It then does a second test for residual noise. The impulse size–based adaptive median filter (SAMF) detects the size of the impulse noise. It then uses the size to adjust the window until it eliminates the noise.

6.1.2 Frame Interpolation

Interpolation is a method for the estimation of new data points from a discrete set of known data points. There

Artificial Neural Networks for Engineering Applications. https://doi.org/10.1016/B978-0-12-818247-5.00015-0

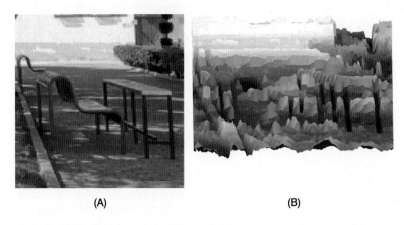

(A)   (B)

FIG. 6.1 **(A)** A grayscale image. **(B)** The signal representing the grayscale image.

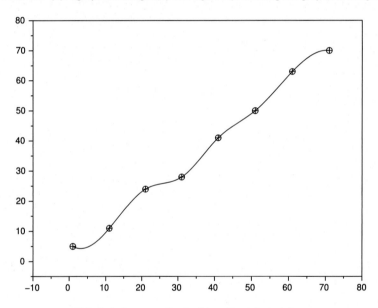

FIG. 6.2 An example of cubic spline interpolation.

are many equations to join these points using interpolation. For linear interpolation the equation is

$$y = y_0 + (x - x_0)\frac{y_1 - y_0}{x_1 - x_0}, \qquad (6.4)$$

where (x_0, y_0) and (x_1, y_1) are the known points and (x, y) is the computed point. In Eq. (6.4), x is the independent variable. For polynomial interpolation the equation is

$$y = \sum_{i=0}^{n} \left(\prod_{0 \le j \le n, j \ne i} \frac{x - x_j}{x_i - x_j} \right) y_i. \qquad (6.5)$$

Interpolation can also use splines. A spline is a piecewise polynomial function. Each polynomial $P_i(x)$ fits $f(x)$ in the interval between x_{i-1} and x_i, i.e.,

$$S(x) = \begin{cases} P_1(x) & x_0 \le x \le x_1 \\ \vdots & \vdots \\ P_i(x) & x_{i-1} \le x \le x_i \\ \vdots & \vdots \\ P_n(x) & x_{n-1} \le x \le x_n \end{cases} \qquad (6.6)$$

In Fig. 6.2 the reader can see an example of a third-degree spline.

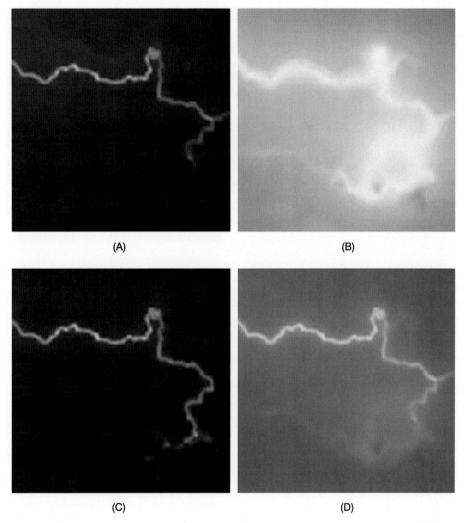

FIG. 6.3 An example using linear frame interpolation. **(A)** Frame 1. **(B)** Frame 3. **(C)** Frame 2. **(D)** is the interpolation of **(A)** and **(B)**.

There exists a method related to interpolation; it is called extrapolation. Extrapolation uses the known data points to predict points outside its range.

One of the main applications of interpolation is increasing the video frame rate. The unit of measure of frame rate is frames per second (fps). Videos create the illusion of movement with the superposition of these frames. The higher the frame rate, the more fluid the video looks. Interpolation can create extra frames with the information in the available frames. These frames do not contain new information, but they increase the superposition. A native higher frame rate will always be better. The reader can observe this in Fig. 6.3. We did

an interpolation between the first and third frame so we can compare it with the second frame. Is easy to see that the two frames are nothing alike and the real frame contains unique information. We also selected the video of a lightning to assert this, because this is a quick phenomenon. In other cases, the difference could not be remarkable. This relates to the Nyquist–Shannon sampling theorem. A quick phenomenon is the equal of a high-frequency signal.

6.1.3 Image Upscaling
In the same manner as increasing the frame rate, interpolation can increase resolution. We call this process

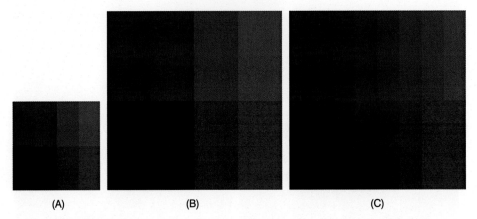

FIG. 6.4 A comparison between a stretched image and an interpolated one. **(A)** Original image. **(B)** Stretched image. **(C)** Interpolated image.

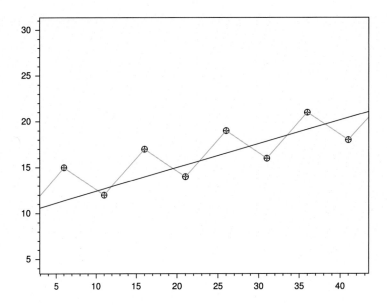

FIG. 6.5 A comparison between linear regression and linear interpolation.

upscaling. This upscaling is not as good as a native high resolution, because the latter will have more details. Even with this in mind, a rescaled image still looks better than a stretched image. From a technical point of view, a stretched image uses a piecewise constant interpolation. In other words, it replicates pixels. This also means from a technical point of view that it has the same resolution, just with bigger pixels. Fig. 6.4 shows an example of this. The interpolation results in a more smooth looking image. And as in the case of frame interpolation, the rescaling cannot add more information.

6.2 SUPPORT VECTOR REGRESSION

Regression is the approximation of points to a function [3]. This is the main difference between regression and interpolation. Interpolation will always pass through the points. Regression will discard the outliers. This can be observed in Fig. 6.5. We choose to compare linear regression and interpolation to emphasize this difference. Another difference is that regression can also predict points outside the range, just like extrapolation. The latter trait could be useful.

Support vector regression (SVR) [10] works with a reformulation for the classification problem. We have the following dual optimization problem:

$$maximize \begin{cases} -\frac{1}{2}\sum_{i,j=1}^{\ell}(\alpha_i - \alpha_i^*)(\alpha_j - \alpha_j^*)\langle x_i, x_j \rangle, \\ -\epsilon \sum_{i=1}^{\ell}(\alpha_i + \alpha_i^*) + \sum_{i=1}^{\ell} y_i(\alpha_i - \alpha_i^*), \end{cases}$$

$$(6.7)$$

subject to

$$\sum_{i=1}^{\ell}(\alpha_i - \alpha_i^*) = 0,$$

$$\alpha_i, \alpha_i^* \in [0, C],$$

$$(6.8)$$

where α and α^* are Lagrange multipliers, ℓ is the number of data points, x are the data points, y are the target values, ϵ and C are user-defined parameters [2], ϵ defines the largest deviation possible from an expected value y, and C regulates the balance between model complexity and the tolerance for values larger than ϵ.

The prediction function is rewritten as

$$f(x) = \sum_{i=1}^{\ell}(\alpha_i - \alpha_i^*)\langle x_i, x \rangle + b,$$

$$(6.9)$$

where b determines the separation of the hyperplane to the origin. The computation of b is as follows:

$$b = \frac{max\{-\epsilon + y_i - \langle \omega, x_i \rangle\} + min\{-\epsilon + y_i - \langle \omega, x_i \rangle\}}{2},$$

$$(6.10)$$

where ω is a linear combination of the training observations, i.e.,

$$\omega = \sum_{i=1}^{\ell}(\alpha_i - \alpha_i^*)x_i.$$

$$(6.11)$$

The operation $\langle x_i, x_j \rangle$ is a dot product. This allows for linear regression. To perform a more complex regression, we need a mapping to higher dimensions. Although this mapping could be explicit, this is not efficient. An implicit mapping with the use of kernels is a better solution. The only condition to use a kernel is that it must fulfill Mercer's condition. If for any $g(x)$

$$\int g(x)^2 dx$$

$$(6.12)$$

is finite, then

$$\iint K(x_i, x_j)g(x_i)g(x_j)dx_i dx_j \geq 0,$$

$$(6.13)$$

and this means that the kernel must be positive semidefinite.

In this work we are going to use the Gaussian kernel

$$gk(x_i, x_j) = e^{-\frac{\|x_i - x_j\|^2}{2\sigma^2}}.$$

$$(6.14)$$

The decision to use the Gaussian kernel comes from two reasons. First, Gaussians are universal function approximators. The generated images will be like the original but still ignoring outliers. The second reason is that the Gaussian kernel is a radial basis function. These kinds of functions only care for the Euclidean distance between the points. This means that the functions are symmetrical and the result is a real number. This allows us to treat each channel independent of the others, combine them, and still get the same result [6]. Otherwise, we must use a quaternion support vector machine.

6.3 SVR METHOD

We next present the method in pseudocode. After that we will proceed to give a detailed explanation. The pseudocode assumes the use of a 0-indexed system. This method executes the three above-mentioned operations, because, as the reader can imagine, training an SVR for each pixel has a high computational cost, but the method computes new pixels by the three tasks using predictions with the trained SVR. The prediction has a tiny cost in comparison, so the best option is to do all the tasks. If the reader wants to do only one, this pseudocode serves as a basis and can be altered with a few changes.

In Algorithm 1 the input data is a video. The output is a video with double the resolution and double the frame rate. We double these to keep the aspect ratio and the running time the same.

In lines 1 and 2 we declare some vectors; tv means training vector, pv means prediction vector, pr means prediction results, and aw is adjustment weights. In line 3 we declare the result video rv.

In line 4 we initialize these vectors and rv. To obtain good results with an SVR we need to normalize the inputs. Taking advantage of this and the way the filtering works, we can define the training vector with 27 positions, going from $(-1, -1, -1)$ to $(1, 1, 1)$ in increments of 1 in each position; pv is similar, going from $(-0.5, -0.5, -0.5)$ to $(0.5, 0.5, 0.5)$ in increments of 0.5

Algorithm 1 SVR for digital video processing.

Data: a video with a $X \times Y$ resolution and Z fps
Result: a video with a $2X \times 2Y$ resolution and $2Z$ fps

1 vector[27][3] pv, tv
2 vector[27] pr, aw
3 rv = video with a $2X \times 2Y$ resolution and $2Z$ fps
4 initialize vectors and rv
5 **for** *x=0 to X-1* **do**
6   **for** *y=0 to Y-1* **do**
7   **for** *z=0 to Z-1* **do**
8   **for** *each channel* **do**
9   **if** *a position of a neighbor is out of bounds* **then**
10   flip the position
11   get y values for position (x,y,z) and its neighbors
12   train SVR model with tv and y
13   pr = prediction of SVR model with pv
14   multiply index x, y, z by 2 and save in nx, ny, nz
15   **for** *i=-1 to 1* **do**
16   **for** *j=-1 to 1* **do**
17   **for** *k=-1 to 1* **do**
18   **if** $i + nx \geq 0$ *and* $j + ny \geq 0$ *and* $k + nz \geq 0$ **then**
19   n = 1
20   **if** *an index is in the edge* **then**
21   multiply n by 2 for each index in the edge
22   compute positions $p1$ and $p2$
23                         $rv(p1)+ = pr(p2) * aw(p2) * n$
24 **return** rv

---

in each position; pr is not initialized. At last the adjustment weights, as the name indicates, are weights used to adjust the contribution of each SVR to 27 pixels in the new video, a $3 \times 3 \times 3$ area. So the number of SVR that contribute to each pixel is

SVR influence

$$
-> \begin{array}{|c|c|c|} \hline 8 & 4 & 8 \\ \hline 4 & 2 & 4 \\ \hline 8 & 4 & 8 \\ \hline \end{array} \begin{array}{|c|c|c|} \hline 4 & 2 & 4 \\ \hline 2 & 1 & 2 \\ \hline 4 & 2 & 4 \\ \hline \end{array} \begin{array}{|c|c|c|} \hline 8 & 4 & 8 \\ \hline 4 & 2 & 4 \\ \hline 8 & 4 & 8 \\ \hline \end{array} ;
$$

$$(6.15)$$

each matrix contributes to a different frame, aw contains the inverse of these coefficients as a vector; rv starts with all its pixels in zero for all channels.

Lines 5 to 8 are the cycles that will go through the whole video and give us the center of each SVR. The if in line 9 is for when we get a position outside the boundaries of the original image. For example, instead of getting the pixel in position $(-1, 2, 3)$, which does not exist, we get $(1, 2, 3)$. This duplication of pixels is prevalent in digital image processing. Another option, because we are using an SVR, is to train without these duplicates. It is necessary to add a term tv for each case. If you are using a precomputed Gramm matrix (GM), you also need one for each case. The if in line 9 will become a more complex selection scheme. The advantage is that the pixels in the edge will be more real. We also will be using the analog of extrapolation for the upper edge pixels. But this only applies to a minuscule subset of pixels in the resulting video. It is most probable that this approach is not worth the effort, but the option is there.

Lines 11 to 13 are the training and prediction steps using the SVR. Since we are using the same tx for all cases, we can precompute a GM to improve the speed of the training.

Line 14 is to get the positions in the new video. Since we are duplicating the dimensions, we multiply the current positions by two.

The cycles at lines 15–17 are for the area of size 27 we mentioned above. The ifs at lines 18 and 20 are there again for the edge pixels of the original video. Negative positions do not exist in the new video, so we discard those in the first if. The second if is for the upper edge pixels in the new video. Each index of the position at this edge means that we have halved the SVRs that determine the new value. So we must multiply the current SVR result by 2 for each one. That is the purpose of the variable n in line 19. Line 22 is self-explanatory; we get the positions based on nx, ny, nz, i, j, k. In line 23 we accumulate the new result with the previous ones.

Finally, we return the result in line 24.

6.4 RESULTS

We did the test presented here with the use of LIBSVM [1] and the Scilab Image and Video Processing (SIVP) Toolbox [7], in conjunction with Scilab.

6.4.1 Image Filtering

The first results will deal with the ability of the filter to remove different types of noise. We did the following modifications to the method to only do image filtering:

FIG. 6.6 Images included in SIVP. **(A)** Baboon. **(B)** Lena. **(C)** Peppers.

- tv is a vector of size [9][2] that goes from $(-1, -1)$ to $(1, 1)$ in increments of 1;
- aw is discarded because each SVR only modifies its central pixel;
- pv is only $(0, 0)$;
- pr is not a vector anymore.

The result is an adaptive Gaussian blur with a windows size of 3×3.

For doing this, for the sake of reproducibility, we will use the images included with SIVP. These are 512×512 color images; see Fig. 6.6. We took the three images and added noise with the function imnoise. We use the default parameters.

We test the quantitative performance through the peak signal-to-noise ratio (PSNR) [9]. The PSNR is

$$PSNR = 10\log_{10}(\frac{255^2}{MSE}), \qquad (6.16)$$

$$MSE = \frac{\sum_i \sum_j \sum_c (r_{ijc} - x_{ijc})^2}{M \times N \times 3}, \qquad (6.17)$$

where r is the original image, x is the filtered image, $M \times N$ is the size of the image, and c represents the value in the RGB channel.

We tested three different types of noise. The first one is Gaussian noise. It has a normal distribution. This noise appears during image acquisition. Some of its causes are low light intensity, high temperature, and transmission. The images are shown in Fig. 6.7.

The second one is impulsive noise, also known as "salt and pepper" noise or spike noise. It appears in the image as white and black dots. Things like faulty memory cells or sensors and errors in broadcast cause this kind of noise. The images are shown in Fig. 6.8.

The last type of noise tested is speckle noise. Speckle noise is granular with a uniform distribution. It is more common in, for instance, radar images. Speckle noise is primarily due to the interference of the returning wave at the transducer aperture. The images are shown in Fig. 6.9.

The SNPRs appear in the following table:

	Baboon	**Lena**	**Peppers**
Gaussian	21.757221	26.614525	26.486678
Impulsive	23.036302	33.614553	32.648606
Speckle	23.064689	31.728948	31.483504

Based on these results, we conclude that the filter is excellent at removing impulsive noise. On the other hand, we also see that Gaussian noise, due its characteristics, is the harder to remove to this filter.

There is one more thing to notice. It can be seen that the noise at the edge of the filtered images has not been removed. We did this on purpose to make it clear why we paid so much attention to it in the explanation of the pseudocode. Without any kind of special consideration, the only option is to ignore the pixels at the edge, and this is the result.

6.4.2 Filtering, Upscaling, and Motion Regression

In this subsection we perform the three tasks with the algorithm. First, we do the same motion compensation with the lightning video, as shown in Fig. 6.10. The limitations with interpolation about new information also appear with regression.

For the comparison between interpolation and regression, we selected two videos. The first one is a CGI animation of space; see Fig. 6.11. The second one is the red card video included in SIVP; see Fig. 6.12.

In Section 6.2, we explained the difference between interpolation/extrapolation and regression. Interpolation will preserve more features of the original image.

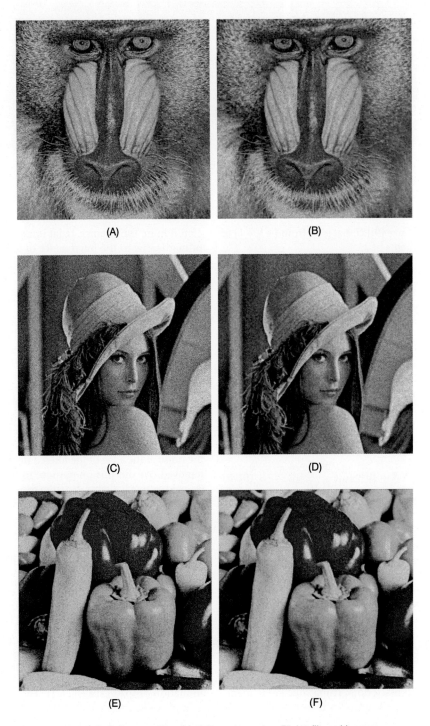

(A)   (B)

(C)   (D)

(E)   (F)

FIG. 6.7 Left: image with added Gaussian noise. Right: filtered images.

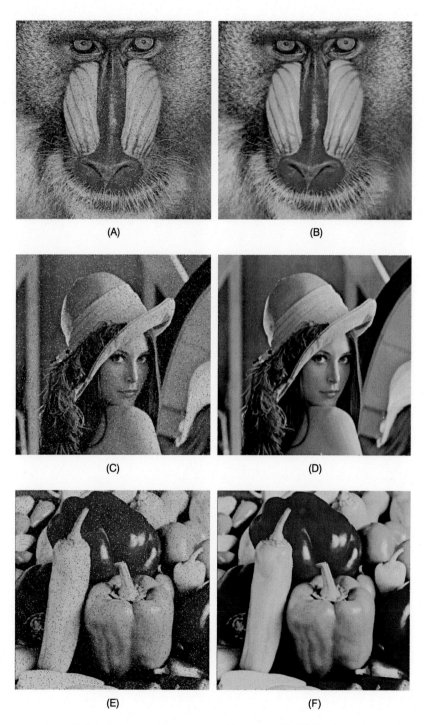

FIG. 6.8 Left: image with added impulsive noise. Right: filtered images.

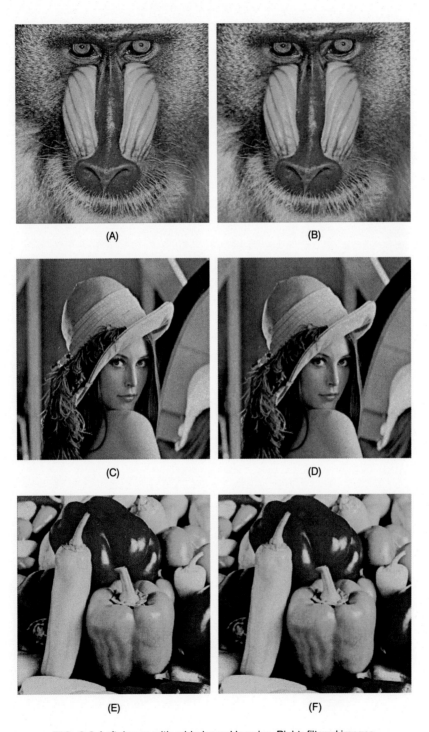

(A)   (B)

(C)   (D)

(E)   (F)

FIG. 6.9 Left: image with added speckle noise. Right: filtered images.

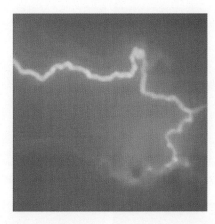

FIG. 6.10 Frame of the lightning video obtained with regression.

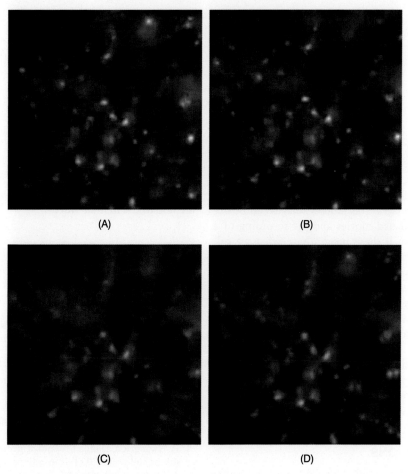

(A)   (B)

(C)   (D)

FIG. 6.11 A comparison between interpolation and regression using artificial images. **(A)** Frame 1.
(B) Frame 2. **(C)** Frame R. **(D)** Frame I.

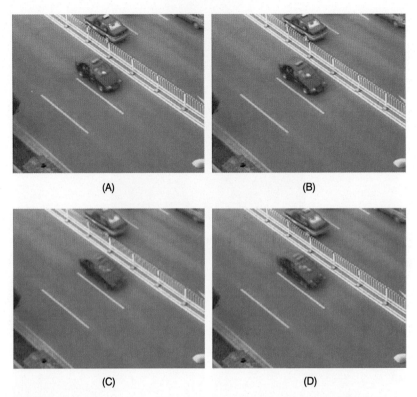

FIG. 6.12 A comparison between interpolation and regression using real images. **(A)** Frame 1. **(B)** Frame 2. **(C)** Frame R. **(D)** Frame I.

Based on this difference, interpolation will perform better on artificial images. This kind of images can be obtained from animated movies and the recording of a video game. Even if these images have what we call outliers, we cannot consider them noise. Its presence is intentional as an artistic choice.

Meanwhile, regression has an inherent filtering process in it. So regression will work better with images coming from cameras. As we explained in Subsection 6.4.1, the sensors in these cameras add noise to the images.

In the CGI video, the reader can observe that the regressed frame looks more blurred than the interpolated one. For the car video the regressed frames carry the flow better than the interpolated one.

6.5 CONCLUSIONS

We have presented a method called SVR for digital video processing. This method is intended for offline video processing and we think it works better with video recordings. It is up to the user to test and decide which method to use in their videos.

REFERENCES

1. C.-C. Chang, C.-J. Lin, LIBSVM: a library for support vector machines, ACM Transactions on Intelligent Systems and Technology 2 (2011) 27.
2. V. Cherkassky, Y. Ma, Selection of meta-parameters for support vector regression, in: J.R. Dorronsoro (Ed.), Artificial Neural Networks — ICANN 2002, Springer Berlin Heidelberg, Berlin, Heidelberg, 2002, pp. 687–693.
3. H. Drucker, C.J.C. Burges, L. Kaufman, A.J. Smola, V. Vapnik, Support vector regression machines, in: M.C. Mozer, M.I. Jordan, T. Petsche (Eds.), Advances in Neural Information Processing Systems 9, MIT Press, 1997, pp. 155–161, http://papers.nips.cc/paper/1238-support-vector-regression-machines.pdf.
4. H. Hwang, R.A. Haddad, Adaptive median filters: new algorithms and results, IEEE Transactions on Image Processing 4 (4) (Apr 1995) 499–502.
5. D.T. Kuan, A.A. Sawchuk, T.C. Strand, P. Chavel, Adaptive noise smoothing filter for images with signal-dependent

noise, IEEE Transactions on Pattern Analysis and Machine Intelligence PAMI-7 (2) (March 1985) 165–177.

6. G. López-González, N. Arana-Daniel, E. Bayro-Corrochano, Parallel Clifford support vector machines using the Gaussian Kernel, Advances in Applied Clifford Algebras 27 (1) (Mar 2017) 647–660, https://doi.org/10.1007/s00006-016-0726-2.

7. S. Savant, SIVP: Scilab image and video processing toolbox 39 (09 2015) 37–38.

8. Scilab Enterprises, Scilab: Free and Open Source software for numerical computation, Scilab Enterprises, Orsay, France, 2012, http://www.scilab.org.

9. I. Singh, N. Neeru, Performance comparison of various image denoising filters under spatial domain 96 (06 2014) 21–30.

10. A.J. Smola, B. Schölkopf, A tutorial on support vector regression, Statistics and Computing 14 (3) (Aug 2004) 199–222, https://doi.org/10.1023/B:STCO.0000035301.49549.88.

11. A. Taguchi, Adaptive α-trimmed mean filters with excellent detail-preserving, in: Proceedings of ICASSP '94. IEEE International Conference on Acoustics, Speech and Signal Processing, vol. v, Apr 1994, pp. 61–64.

CHAPTER 7

Artificial Neural Networks Based on Nonlinear Bioprocess Models for Predicting Wastewater Organic Compounds and Biofuel Production

KELLY J. GURUBEL, DSC • EDGAR N. SANCHEZ, DSC •
RAFAEL GONZÁLEZ, MENG • HUGO COSS Y LEÓN, MSC •
ROXANA RECIO, MSC

7.1 INTRODUCTION

The rapid increase in wastewater production due to domestic, industrial, and agricultural activities requires careful consideration by all societal sectors. Wastewater treatment (WWT) is necessary to reduce the level of organic material and suspended solid pollutants [38]. The most adequate biological methods for WWT are activated sludge and anaerobic digestion. They provide a wide variety of advantages, including environmental benefits, as well as economic ones [9]. Nevertheless, selecting appropriate WWT alternatives, from an integrative perspective, is difficult and complex, given the variety of linked objectives and conflicting criteria that must be considered [45]. WWT biological processes are very attractive due to their waste treatment properties and their capacity for generating biofuel from waste materials, which can be used for electrical energy generation [23,26]. However, the processes are sensible to variations in operating conditions, such as pH, temperature, hydraulic and organic overloads, and aeration [7,37]. In biological processes, variable measurements are hardly measurable and relatively expensive or time consuming, so process supervision and control are necessary [22]. Therefore, estimation strategies are essential to the identification of variables that are difficult to access, for monitoring and control applications. The adaptive observer [36], for nonlinear cascade state affine systems, gives a reliable tool towards the control design considering estimated states and parameters of the model. Based on sensitivity analysis, the proposed nonlinear adaptive observer is able to estimate the most sensitive parameters and system states with an arbitrarily tunable rate. This observer is tested for acidogenic and methanogenic bacteria concentration estimations, as well as growth rate parameter estimations for an anaerobic digestion process. This scheme shows an adequate performance in asymptotic convergence, even under uncertain kinetic parameter variations and noisy measurements; nevertheless, convergence is briefly lost for abrupt changes in dilution rate. The reset adaptive observer (ReAO) for nonlinear systems is an adaptive observer consisting of an integrator and a reset law that resets the output of the integrator depending on a predefined reset condition [1]. The inclusion of reset elements can improve the observer performance but it can also destroy the stability of the estimation process if the ReAO is not properly tuned. The observer gains and the reset element parameters are optimally chosen by solving the L_2 gain minimization problem, which can be rewritten as an equivalent linear matrix inequality (LMI) problem. The two most popular reset conditions within the reset time–independent framework are zero crossing and sector condition. The main advantage of ReAOs is that potentially much richer feedback signals can be obtained by resetting some observer states. Since the method is based on minimizing the L_2 gain of the ReAO, the stability and convergence of the estimation process are guaranteed. An exception would be when the system is affected by disturbances with a steady-state offset, since the ReAO might need more time to reject the effect of the disturbance on the unknown variables. Artificial intelligence (AI)-based observers perform the estimation task coupled with conventional (model-based) techniques by using rule-based methods such as fuzzy logic, artificial neural networks, and genetic algorithms [29,31]. In [30], a hybrid observer is used on the ethylene polymerization process to estimate the ethylene and butene concentrations in the reactor as well as the melt flow index. The proposed hybrid observer provides fast estimation with a high rate of accuracy even in the

Artificial Neural Networks for Engineering Applications. https://doi.org/10.1016/B978-0-12-818247-5.00016-2

presence of disturbances and noise in the model. In [34], the biogas production process in a batch biodigester was modeled with a hybrid artificial neural network. The best combination of the substrates for maximum biogas production was established by coupling the ANN model developed with a genetic algorithm. The coupled scheme is an efficient optimizing tool for the biogas production process. The formulation of an AI-based observer may be difficult and time consuming compared to the other hybrid observers in some systems. Discrete-time nonlinear Recurrent High-Order Neural Observer (RHONO) for unknown nonlinear systems in the presence of external disturbances and parameter uncertainties is proposed in [3]. This neural observer based on the recurrent high-order neural network (RHONN) has proven to be effective for biological processes [2,25]. Recurrent neural networks have at least one feedback loop, which improves the learning capability and performance of the network. This structure also offers a better-suited tool to model and control nonlinear systems [5]. The main advantages that this neural network offers are their high performance and its low level of complexity and tuning; additionally, knowledge of the model is not strictly necessary.

The anaerobic digestion process has received increasing attention due to its potential for generating renewable energy from several organic compounds [24]. In the literature, recent studies have demonstrated the feasibility of alternative substrates for bioenergy production through anaerobic digestion [40,32,8]. Two-stage anaerobic digestion for simultaneous hydrogen and methane production is considered to be more efficient in biofuel yield, energy recovery rate, and chemical oxygen demand (COD) [16]. During the process, the acidogenesis and methanogenesis stages operate separately, which increases the stability by controlling the acidification phase in the hydrogen producing stage and hence preventing the inhibition of the methanogenic population during methane production [42]. Anaerobic digestion models for the continuous two-stage process for specific substrates have been reported [44,21,6]. The two independent processes enable optimal conditions for the acidogenic and the methanogenic bacterial biomass to be established; however, the model is not ideal for each application, so specific extensions must be added for a better process estimation [10,48,27]. Reduced models have been proposed that reproduce the most relevant phenomena of the anaerobic digestion process for biofuel production; nevertheless, there are still nonlinear states that are hard to measure [19,20,13]. The modeling and simulation trends of the anaerobic biological reactors focus on structure and

extension improvements; however, due to instability in the anaerobic digestion operation, it is necessary to propose control strategies that contribute to efficiency in the reduction of contaminants and inhibitors, the conversion of intermediate compounds, and continuous biofuel production. In order to estimate the process and the continuous biofuel production, artificial neural networks have been proposed for complex dynamics estimation and control strategies to increase biogas yield [28,11,15]. There are different nonlinear neural model structures for system identification, like neural networks with random weights (NNRWs), neuro-fuzzy models (NFMs), and multilayer perceptron (MLP) neural networks, and hybrid dynamic neural network structures [46,14,17]. The difficulties of these structures lie in their low flexibility, robustness and lack of affinity for optimal control design. The recurrent neural network with external input (NNARX) structure consists of exogenous inputs, which means that the model relates the current value of a time series to past values of that series, or to the present and past values of the same series for predictions. The NNARX network is one of the dynamic networks which have been used in different fields for time series prediction. It has been demonstrated that NNARX recurrent neural networks have the potential to capture the dynamics of nonlinear complex systems [18]. The main motivation for using this neural structure in system identification is to obtain a model with physical meaning, which is not possible with feedforward neural networks. In the neural model, the state of each neuron represents a state variable of the system to be modeled, which means a considerable reduction in the number of neurons in the network [4]. The structure trained with past values of the output allows to find a stable prediction of real-value time series. In this chapter, a RHONO structure to estimate dynamic states related to the output effluent quality based on COD in an anaerobic wastewater treatment model is proposed. The neural observer performance to estimate the organic compounds reduction at the system output is evaluated in the presence of disturbances. Thereafter, a NNARX model structure for a two-stage anaerobic digestion model with simultaneous hydrogen and methane production in the presence of external disturbances is proposed. The neural network identifies the complex dynamics of a reduced model derived from the anaerobic digestion model 1 (ADM1), which describes the process of anaerobic digestion of organic waste in two continuously stirred reactors. The proposed neural network structures facilitate the availability of the significant states of the process in order to estimate the organic

TABLE 7.1
State variable descriptions.

Parameter	Unit	Description
$Z_{1,k}$	mg COD L^{-1}	Readily biodegradable soluble substrate
$Z_{2,k}$	mg COD L^{-1}	Slowly biodegradable soluble substrate
$Z_{3,k}$	mg L^{-1}	Soluble oxygen
$Z_{4,k}$	mg COD L^{-1}	Active heterotrophic particulate biomass
$Z_{5,k}$	mg COD L^{-1}	Active autotrophic particulate biomass
$Z_{6,k}$	mg N L^{-1}	Soluble nitrate and nitrite nitrogen
$Z_{7,k}$	mg N L^{-1}	Soluble ammonium nitrogen
$Z_{8,k}$	mg N L^{-1}	Soluble biodegradable organic nitrogen
$Z_{9,k}$	mg COD L^{-1}	Particulate biodegradable organic nitrogen

TABLE 7.2
Reaction rates.

Reaction rate	Unit	Description
M_2	–	Monod kinetics for readily biodegradable
M_{8a}	–	Monod kinetics for the component $Z_{3,k}$ with respect autotrophic biomass
M_{8h}	–	Monod kinetics for the component $Z_{3,k}$ with respect to heterotrophic biomass
M_9	–	Monod kinetics for soluble nitrate and nitrite nitrogen
M_{10}	–	Monod kinetics for soluble ammonium nitrogen
I_8	–	Inhibition kinetics for soluble oxygen
K_{sat}	–	Saturation kinetics

compounds reduction and to predict biofuel production.

7.2 ACTIVATED SLUDGE PROCESS

7.2.1 Activated Sludge Model

The activated sludge model 1 (ASM1) consists of 13 nonlinear ordinary differential equations, composed of seven soluble organic matter and six suspended organic matter components [43]. The model describes soluble and particulate carbonaceous components, soluble and particulate nitrogen components, soluble biodegradable organic nitrogen, and particulate biodegradable organic nitrogen. In this work, a reduced model derived from ASM1 describing the aerobic digestion process of organic wastewater in a continuously stirred reactor is considered [33]. The following assumptions are made: (a) the reactor is well mixed and the hydrolysis rate of soluble biomass obtained from the wastewater is the same as the hydrolysis rate of biomass obtained from

the degradation of biomass; (b) four dynamic states pertaining to inert organic material and nonbiodegradable particulate products are uncoupled to simplify the model. The reduced model consists of nine nonlinear ordinary differential equations, which describe the present processes in the reactor. The discrete-time model is represented by Eqs. (7.1)–(7.9); state variables are described in Table 7.1 and the reaction rates in Table 7.2. We have

$$Z_{1,k+1} = Z_{1,k} + T\left(\frac{q}{V}\left(Z_{1,in} - Z_{1,k}\right)\right.$$
$$\left. - \frac{1}{YH}\cdot\mu_{max,H}M_2(M_{8h} + I_8M_9n_g)Z_{4,k} \quad (7.1)\right.$$
$$\left. + k_h k_{sat}\left(M_{8h} + n_h I_8 M_9\right)Z_{4,k}\right),$$

$$Z_{2,k+1} = Z_{2,k} + T\left(\frac{q}{V}\left(Z_{2,in} - Z_{2,k}\right) + \frac{qr}{V}(b-1)Z_{2,k}\right.$$
$$\left. + (1 - fp)(b_H Z_{4,k} + b_A Z_{5,k}) \quad (7.2)\right.$$
$$\left. - k_h\cdot k_{sat}(M_{8H} + n_h\cdot I_8\cdot M_9)Z_{4,k}\right),$$

$$Z_{3,k+1} = Z_{3,k} + T\left(\tfrac{q}{V}\left(Z_{3,in} - Z_{3,k}\right)\right.$$
$$+ K_{L,A}\left(Z_{3,\max} - Z_{3,k}\right)$$
$$- \tfrac{(1-Y_H)}{Y_H}\mu_{\max H} M_2 M_{8H} Z_{4,k}$$
$$\left. - \tfrac{(4.57 - Y_A)}{Y_A}\mu_{\max,A} M_{10} M_{8a} Z_{4,k}\right),$$
(7.3)

$$Z_{4,k+1} = Z_{4,k} + T\left(\tfrac{q}{V}\left(Z_{4,in} - Z_{4,k}\right)\right.$$
$$+ \tfrac{qr}{V}(b-1) Z_{4,k} + \mu_{\max,H} M_2 M_{8h} X_{B,H}$$
$$\left. + \mu_{\max,H} M_2 I_8 M_9 ng\, Z_{4,k} - b_H Z_{4,k}\right),$$
(7.4)

$$Z_{5,k+1} = Z_{5,k} + T\left(\tfrac{q}{V}\left(Z_{5,in} - Z_{5,k}\right) + \tfrac{qr}{V}(b-1) Z_{5,k}\right.$$
$$\left. + \mu_{MAX,A} M_{10} M_{8a} Z_{5,k} - b_A Z_{5,k}\right),$$
(7.5)

$$Z_{6,k+1} = Z_{6,k} + T\left(\tfrac{q}{V}\left(Z_{6,in} - Z_{6,k}\right)\right.$$
$$- \tfrac{(1-Y_H)}{2.86 Y_H}\mu_{\max H} M_2 I_8 M_9 n_g Z_{4,k}$$
$$\left. + \tfrac{1}{Y_A}\mu_{MAX,A} M_{10} M_{8a} Z_{5,k}\right),$$
(7.6)

$$Z_{7,k+1} = Z_{7,k} + T\left(\tfrac{q}{V}\left(Z_{7,in} - Z_{7,k}\right)\right.$$
$$- i_{XB}\mu_{\max,H}.M_2\left(M_{8h} + I_8.M_9.ng\right) Z_{4,k}$$
$$- \left(i_{XB} + \tfrac{1}{Y_A}\right)\mu_{\max,A}.M_{10}.M_{8a}.Z_{5,k}$$
$$\left. + k_A.S_{ND}.Z_{4,k}\right),$$
(7.7)

$$Z_{8,k+1} = Z_{8,k} + T\left(\tfrac{q}{V}\left(Z_{8,in} - Z_{8,k}\right) - k_A.Z_{8,k}.Z_{4,k}\right.$$
$$\left. + k_h k_{sat}\left(M_{8h} + n_h I_8 M_9\right) Z_{4,k}\tfrac{Z_{9,k}}{Z_{2,k}}\right),$$
(7.8)

$$Z_{9,k+1} = Z_{9,k} + T\left(\tfrac{q}{V}\left(Z_{9,in} - Z_{9,k}\right) + \tfrac{qr}{V}(b-1) Z_{9,k}\right.$$
$$+ \left(i_{XB} - f_p i_{XP}\right)\left(b_H Z_{4,k} + b_A Z_{5,k}\right)$$
$$\left. - k_h k_{sat}\left(M_{8h} + n_h I_8 M_9\right) Z_{4,k}\tfrac{Z_{9,k}}{Z_{2,k}}\right).$$
(7.9)

The corresponding reaction rates are

$$M_2 = \frac{Z_{1,k}}{K_S + Z_{1,k}},$$
(7.10)

$$M_{8a} = \frac{Z_{3,k}}{K_{O,A} + Z_{3,k}},$$
(7.11)

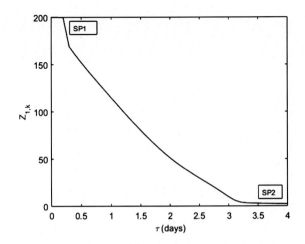

FIG. 7.1   Readily biodegradable soluble substrate vs HRT.

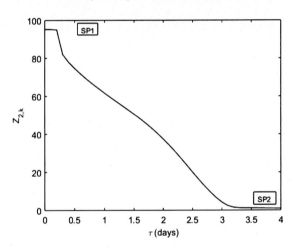

FIG. 7.2   Slowly biodegradable particulate substrate vs HRT.

$$M_{8h} = \frac{Z_{3,k}}{K_{O,H} + Z_{3,k}},$$
(7.12)

$$M_9 = \frac{Z_{6,k}}{K_{NO} + Z_{6,k}},$$
(7.13)

$$M_{10} = \frac{Z_{7,k}}{K_{NH} + Z_{7,k}},$$
(7.14)

$$I_8 = \frac{K_{O,H}}{K_{O,H} + Z_{3,k}},$$
(7.15)

$$k_{sat} = \frac{Z_{2,k}}{K_X Z_{4,k} + Z_{2,k}}.$$
(7.16)

The steady state for the readily biodegradable soluble substrate and the slowly biodegradable particulate substrate are shown in Figs. 7.1 and 7.2.

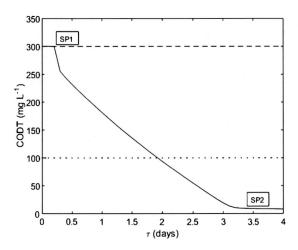

FIG. 7.3 Total chemical oxygen demand (CODT) vs HRT.

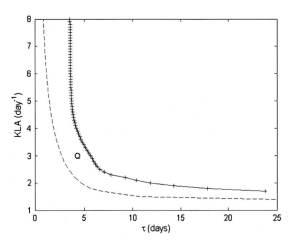

FIG. 7.4 Classification of ASM performance as a function of the KLA.

These diagrams are related to the total COD at the output system. Two stability sections are displayed in these diagrams, i.e., sections SP1 and SP2. In Fig. 7.1, the first section, SP1, at a hydraulic residence time (HRT) of 0.2 days, determines the condition under which the wash-out solution loses stability. Wash-out occurs when the biological reactor cannot support any microorganisms. Thus, the HRT must be greater than this value to prevent wash-out occurring in the system. On the other hand, section SP2 shows the reached stability of the substrate at an HRT of 3.4 days. In the second section, SP2, the substrate concentration decreases marginally with increasing values of the HRT. Between the first and second sections, the value of the biodegradable soluble substrate decreases steadily with increasing HRT to reach the value of 3.2 mg L^{-1}. At this point, the substrate reduction is 98.4% of the concentration entering the reactor. Similar results can be concluded from Fig. 7.2.

The performance of wastewater treatment plants is measured as the effluent quality. For this work, the allowable concentrations of COD leaving a wastewater treatment plant specified by Mexican legislation are considered. The Mexican legislation establishes a maximum value of 100 mg L^{-1} of COD [41].

Thus, we quantify effluent quality using COD as the performance index. The objective is to determine the residence time such that the total COD (CODT) $\leq$100 mg L^{-1}. The CODT is defined in (7.28).

Fig. 7.3 shows the CODT steady state as a function of the HRT. As before, the CODT decreases as the HRT increases from the value at section SP1, towards the value at section SP2. The decrease is marginal for HRT

values greater than section SP2. In order to satisfy the legislative requirement of CODT $=$100 mg L^{-1} (dotted horizontal line), the HRT must be chosen such that the CODT is below this value. To accomplish the Mexican legislation, the biological reactor must be operated at a residence time longer than 1.9 days. Thus operating the reactor at a residence time of 3.4 days ensures that the CODT value in the effluent is only around 8% of the legislative requirement.

According to this analysis, if the aerobic sludge process performance is to be significantly improved with very large residence times, a different reactor configuration, such as a cascade, is recommended.

From the automatic control point of view, the control of activated sludge reactors is often based on the oxygen transfer coefficient (KLA). In order to explain the ASM process dependence upon the choice of KLA, Fig. 7.4 shows the values of the HRT as a function of the KLA. The line through the asterisks represents section points at which increasing the HRT only increases the performance marginally. The dashed line is the value of the HRT at which legislative requirements are met. To accomplish the Mexican CODT legislation, from an operational point of view, the values of KLA and HRT must be in the Q region.

7.2.2 Discrete-Time RHONO

A discrete-time nonlinear RHONO for unknown nonlinear systems in the presence of external disturbances and parameter uncertainties is described in [4]. This observer is based on a discrete-time RHONN trained with an extended Kalman filter (EKF)-based algorithm. First, let us consider the following nonlinear system, which is assumed to be observable

$$x_{k+1} = F(x_k, u_k) + d_k,$$
$$y_k = h(x_k),$$
(7.17)

where $x_{k+1} \in R^n$ is the state vector of the system, $u_k \in R^m$ is the input vector, $h(x_k) \in R^{p \times n}$ is a nonlinear function of the system states, $d_k \in R^n$ is a disturbance vector, and $F(\bullet)$ is a smooth vector field; hence (7.17) can be also expressed componentwise as

$$x_{i,k+1} = F_i(x_k, u_k) + d_{i,k}, \qquad i = 1, ..., n,$$
$$x_k = [x_{1,k}...x_{i,k}...x_{n,k}]^T,$$
$$d_k = [d_{1,k}...d_{i,k}...d_{n,k}]^T,$$
$$y_k = h(x_k).$$
(7.18)

For system (7.18), a Luenberger-like neural observer is proposed, with the following structure:

$$\hat{x}_{i,k+1} = w_i^T z_i(\hat{x}_k, u_k) + g_{mi}e_k,$$
$$\hat{x}_k = [\hat{x}_{1,k}...\hat{x}_{i,k}...\hat{x}_{n,k}]^T,$$
$$\hat{y}_k = h(\hat{x}_k); \qquad i = 1, ..., n,$$
(7.19)

where $g_{mi} \in R^p$ is the Luenberger gain, u_i is the external input vector to the neural network, and z_i is a function of states and inputs to each neuron; the weight vectors are updated online with a decoupled EKF. The output error is defined as

$$e_k = y_k - \hat{y}_k.$$
(7.20)

The weight estimation error is defined as

$$\tilde{w}_{i,k} = w_{i,k} - w_i*,$$
(7.21)

where w_i* is the ideal weight vector and w_i its estimate. The general discrete-time nonlinear system (7.17), which is assumed to be observable, can be approximated by the following discrete-time RHONN parallel representation:

$$x_{i,k+1} = w_i^{*T} z_i(x_k, u_k) + \epsilon_{z_i}, \quad i = 1, ..., n, \quad (7.22)$$

where ϵ_{z_i} is a bounded approximation error, which can be reduced by increasing the number of the adjustable weights. Assume that there exists an ideal unknown weights vector w_i* such that $\| \epsilon_{z_i} \|$ can be minimized on a compact set $\Omega z_i \subset R^{Li}$. The ideal weight vector w_i* is an artificial quantity required only for analytical purposes and is defined as

$$w_i^* = \arg\min_{w_i} \left\{ \sup_{x,u} \left| F_i(x_k, u_k) - w_i^T z_i(\bullet) \right| \right\}, \quad (7.23)$$

which is assumed to be unknown, and it constitutes the optimal set which renders the minimum approximation error, defined as ϵ_{z_i}; $F_i(\bullet)$ is the ith component of $F(\bullet)$ [47]. EKF-based algorithms have been introduced to train neural networks, improving learning convergence. Since the neural network mapping is nonlinear, an EKF type is required. The training goal is to find the optimal weight values, which minimize the predictions error. More details are presented in [39,35]. An EKF-based training algorithm is described as follows:

$$w_{i,k+1} = w_{i,k} + \eta_i K_{i,k} e_{i,k},$$
$$K_{i,k} = P_{i,k} H_{i,k} M_{i,k},$$
$$P_{i,k+1} = P_{i,k} - K_{i,k} H_{i,k}^T P_{i,k} + Q_{i,k},$$
$$i = 1, ..., n,$$
(7.24)

with

$$M_{i,k} = \left[R_{i,k} + H_{i,k}^T P_{i,k} H_{i,k} \right]^{-1},$$
$$e_{i,k} = y_k - \hat{y}_k,$$
(7.25)

where $e_{i,k} \in R^P$ is the observation error, $P_{i,k} \in R^{Li \times Li}$ is the state prediction error–associated covariance matrix at step k, $w_{i,k} \in R^{Li}$ is the weight (state) vector, L_i is the total number of neural network weights, $y \in R^P$ is the measured output vector, $\hat{y} \in R^P$ is the neural network output, η_i is the learning rate parameter, $K_{i,k} \in R^{Li \times P}$ is the Kalman gain matrix, $Q_{i,k} \in R^{Li \times Li}$ is the state noise–associated covariance matrix, $R_{i,k} \in R^{P \times P}$ is the measurement noise–associated covariance matrix, and $H_{i,k} \in R^{Li \times P}$ is a matrix for which each entry (H_{ij}) is the derivative of the ith neural output with respect to the ijth neural network weight, (w_{ij}), where $i = 1, ..., n$ and $j = 1, ..., L_i$. Usually P_i, Q_i, and R_i are initialized as diagonal matrices, with entries $P_{i,0}$, $Q_{i,0}$, and $R_{i,0}$ respectively. It is important to remark that $H_{i,k}$, $K_{i,k}$, and $P_{i,k}$ for the EKF are bounded [39].

The output error e_k is defined as

$$e_k = y_k - \hat{y}_k.$$
(7.26)

7.2.3 Neural Observer Structure

The proposed RHONO estimates the readily biodegradable soluble substrate, the slowly biodegradable particulate substrate, and the soluble oxygen dynamics related to COD, which is commonly measured in this process. The observability property of the ASM was analyzed in a previous work [12]. The convergence is tested at random initial conditions with a constant HRT $\tau = 3.4$ days and

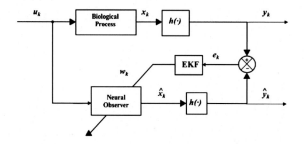

FIG. 7.5 Neuronal identification scheme.

a transfer oxygen coefficient KLA = 4 day^{-1}. The neural network structure is shown in (7.27), the proposed neural observer has a parallel configuration, and the vector of weights w_{ij} is updated online with the EKF given by (7.24)–(7.25). The sampling time is one day and the RHONO scheme is illustrated in Fig. 7.5. We have

$$\hat{Z}_{1,k+1} = w_{11,k} S\left(\hat{Z}_{1,k}\right) + w_{12,k} S^2\left(\hat{Z}_{1,k}\right)$$
$$+ w_{13,k} S\left(\hat{Z}_{2,k}\right) + w_{14,k} S^2\left(\hat{Z}_{1,k}\right)$$
$$+ w_{15,k} S^2\left(\hat{Z}_{2,k}\right) KLA_k + g_1 e_k,$$

$$\hat{Z}_{2,k+1} = w_{21,k} S\left(\hat{Z}_{2,k}\right) + w_{22,k} S^2\left(\hat{Z}_{2,k}\right)$$
$$+ w_{23,k} S\left(\hat{Z}_{1,k}\right) + w_{24,k} S^2\left(\hat{Z}_{2,k}\right) \quad (7.27)$$
$$+ w_{25,k} S^2\left(\hat{Z}_{3,k}\right) KLA_k + g_2 e_k,$$

$$\hat{Z}_{3,k+1} = w_{31,k} S\left(\hat{Z}_{3,k}\right) + w_{32,k} S^2\left(\hat{Z}_{3,k}\right)$$
$$+ w_{33,k} S\left(\hat{Z}_{1,k}\right) + w_{34,k} S^2\left(\hat{Z}_{3,k}\right)$$
$$+ w_{35,k} S^2\left(\hat{Z}_{3,k}\right) KLA_k + g_3 e_k,$$

where w_{ij} is the respective online adapted weight vector; $\hat{Z}_{1,k}$ is the readily biodegradable soluble substrate (mg COD L^{-1}), $\hat{Z}_{2,k}$ is the slowly biodegradable particulate substrate (mg COD L^{-1}), $\hat{Z}_{3,k}$ the soluble oxygen (mg L^{-1}), $S(\bullet)$ is the sigmoid function defined as $S(x) = \alpha \tanh(\beta x)$, (g_1, g_2, g_3) are the Luenberger-like observer gains, and e_k is the output error.

The observer output is defined as follows:

$$\hat{y}_k = \hat{Z}_{1,k} + Z_{1,I} + \hat{Z}_{2,k} + Z_{2,I}, \quad (7.28)$$

where $Z_{1,I}$ and $Z_{2,I}$ are the inert soluble organic material and the particulate inert organic matter, respectively.

This scheme illustrates the process input $u_k \in R^m$, $x_k \in R^n$ represents the states of the system, $\hat{x}_k \in R^n$ represents the estimated states, $y_k \in R^p$ represents the measurable output of the system, $\hat{y}_k \in R^p$ is the estimated output of the observer, and e_k is the output error. The

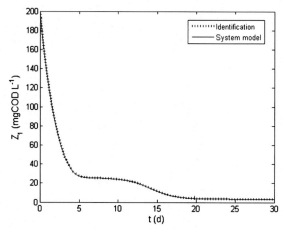

FIG. 7.6 Readily biodegradable soluble substrate estimation.

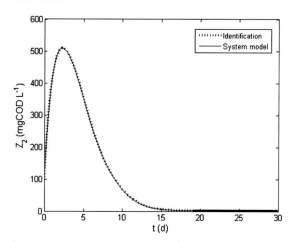

FIG. 7.7 Slowly biodegradable particulate substrate estimation.

performance of the RHONO is validated via simulation and shown in Figs. 7.6–7.8.

As can be seen, ASM dynamic states are well estimated along simulation time with KLA as step input and random initial conditions. The neural observer presents a fast convergence with a good performance along simulation time with neglected transient state error. Thus, the proposed neural observer trained with an EKF-based algorithm is a good alternative to estimate those complex states.

7.2.4 RHONO Performance in the Presence of Disturbances

In order to show the observer performance in the presence of input disturbances, an analysis with time-

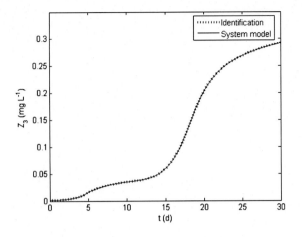

FIG. 7.8 Soluble oxygen estimation.

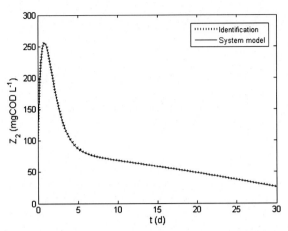

FIG. 7.10 Z_2 estimation with time-varying KLA and $\tau = 1$.

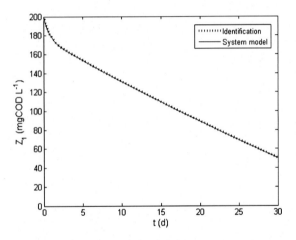

FIG. 7.9 Z_1 estimation with time-varying KLA and $\tau = 1$.

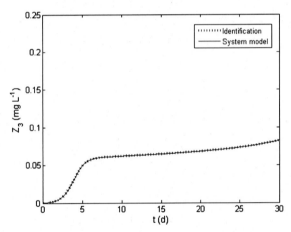

FIG. 7.11 Z_3 estimation with time-varying KLA and $\tau = 1$.

varying KLA is realized. First, convergence is tested with random initial conditions, a constant HRT of $\tau = 1$ day, and a time-varying KLA from 1 to 8 day^{-1} for 30 days. Results are displayed in Figs. 7.9–7.11.

According to Fig. 7.4, the legislative COD concentration leaving the wastewater treatment plant can be accomplished considering the reactor configuration, KLA, and HRT. In order to test the neural performance, KLA is selected as time-varying and HRT is fixed as constant. As can be seen in Figs. 7.9–7.11, the neural observer has a good performance estimation for all states in the presence of disturbances; Z_1 and Z_2 substrate concentrations rapidly decrease as the KLA increases. Excess of soluble oxygen is displayed in Fig. 7.11. This test shows that the legislative COD concentration is met with the proposed operating conditions. Nevertheless,

several operating conditions can be determined by selecting different values of HRT and KLA.

As a second test, convergence is analyzed with random initial conditions, a constant $\tau = 6$ days, and a time-varying KLA from 1 to 8 day^{-1} for 30 days.

Figs. 7.12–7.14 display the neural observer performance in the presence of disturbances; Z_1 and Z_2 substrate concentrations are well estimated and the steady state is reached at 15 days. Excess of soluble oxygen is displayed in Fig. 7.14. The new HRT contributes to a rapid decrease of the substrate concentrations with the same time-varying KLA value. This test shows that the legislative COD concentration is met with the proposed operating conditions. Nevertheless, for this operating point reactor configuration must be considered.

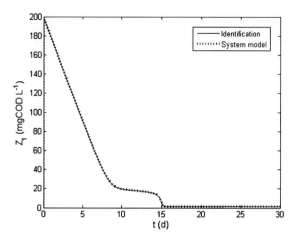

FIG. 7.12 Z_1 estimation with time-varying KLA and $\tau = 6$.

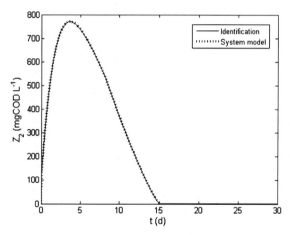

FIG. 7.13 Z_2 estimation with time-varying KLA and $\tau = 6$.

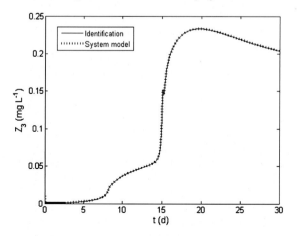

FIG. 7.14 Z_3 estimation with time-varying KLA and $\tau = 6$.

TABLE 7.3
State variables of the model.

State	Unit	Description
x_1	g L^{-1}	Acidogenic bacteria
x_2	g L^{-1}	Acetogenic bacteria
x_3	g L^{-1}	Methanogenic bacteria
s_1	g COD L^{-1}	Monosaccharides
s_2	g COD L^{-1}	Amino acids
s_3	g COD L^{-1}	Fatty acids
s_4	g COD L^{-1}	Total acetate
s_5	g COD L^{-1}	Composites
s_6	g COD L^{-1}	Carbohydrates
s_7	g COD L^{-1}	Proteins
s_8	g COD L^{-1}	Lipids
s_9	g COD L^{-1}	Acetate

7.3 ANAEROBIC DIGESTION PROCESS

7.3.1 Two-Stage Anaerobic Digestion Model

The reduced model is derived from ADM1 and describes the anaerobic digestion process of organic waste in two continuously stirred reactors for simultaneous hydrogen and methane production [13]. The following assumptions are made: (a) the slow acetogenesis phase is omitted and consequently in the second reactor only the methanogenesis phase exists; (b) balance equations of the hydrogen and methane in the liquid phases are disregarded because they are not dissolved in liquids; (c) the hydrogenotrophic bacteria are not present in the process; (d) the pH is kept constant in both reactors; and (e) the balance of inorganic components and some biochemical equations have been neglected to simplify the model. The reduced model consists of twelve ordinary differential equations, which describe the presented processes in both reactors. The state variables which represent the concentrations are described in Table 7.3. We have

$$\dot{x}_1 = (-D_1 + Y_{sa}\mu_{1sa})x_1 + Y_{sa}\mu_{2sa}x_1, \qquad (7.29)$$

$$\dot{x}_2 = (-D_1 + Y_f\mu_{3f})x_2, \qquad (7.30)$$

$$\dot{x}_3 = (-D_2 + Y_{ch}\mu_{9ac})x_3, \qquad (7.31)$$

$$\dot{S}_1 = D_1(S_{1in} - S_1) + k_{hc}S_6 + f_{1l}k_{hl}S_8 - \mu_{1sa}x_1, \qquad (7.32)$$

$$\dot{S}_2 = D_1(S_{2in} - S_2) + k_{hp}S_7 - \mu_{2sa}x_1, \qquad (7.33)$$

$$\dot{S}_3 = D_1(S_{3in} - S_3) + f_{3l}k_{hl}S_8 - \mu_{3f}x_2, \qquad (7.34)$$

TABLE 7.4
Parameters of the model.

Parameter	Unit	Description
$S_{1in}, S_{2in}, S_{3in}, S_{5in}$	g COD L^{-1}	Acidogenic bacteria
D_1, D_2	h^{-1}	Dilution rates
$f_{cx}, f_{px}, f_{lx}, f_{1l}, f_{3l}, f_{4s}, f_{4a}, Y_{sa}, Y_{ac}, Y_f$		Stoichiometric parameters
$Y_{h2s}, Y_{h2a}, Y_{h2f}, Y_{ch}$	L^2 g^{-1}	Physicochemical parameters
$k_d, k_{ch}, k_{hp}, k_{hl}$	h^{-1}	Biochemical parameters

$$\dot{S}_4 = -D_1 S_4 + (1 - Y_{sa}) f_{4s}\mu_{1sa}x_1$$
$$+ (1 - Y_{sa}) f_{4a}\mu_{1sa}x_1 + 0.7(1 - Y_f)\mu_{3f}x_2, \tag{7.35}$$

$$\dot{S}_5 = D_1(S_{5in} - S_5) - k_d S_5, \tag{7.36}$$

$$\dot{S}_6 = -(D_1 + k_{hc})S_6 + f_{cx}k_d S_5, \tag{7.37}$$

$$\dot{S}_7 = -(D_1 + k_{hp})S_7 + f_{px}k_d S_5, \tag{7.38}$$

$$\dot{S}_8 = -(D_1 + k_{hl})S_8 + f_{lx}k_d S_5, \tag{7.39}$$

$$\dot{S}_9 = D_2(S_4 - S_9) - Y_{ac}\mu_{9ac}x_3. \tag{7.40}$$

The stoichiometric and kinetic parameters are defined in Table 7.4. The gaseous output is defined as follows:

$$Q_{h_2} = (Y_{h2s}\mu_{1sa} + Y_{h2a}\mu_{2sa})x_1 + Y_{h2f}\mu_f x_2, \tag{7.41}$$

$$Q_{ch_4} = Y_{ch}\mu_{9ac}x_3, \tag{7.42}$$

where Q_{h_2} is the hydrogen gaseous stream in the first reactor and Q_{ch_4} the methane gaseous stream in the second reactor. The hydrogen and methane equations are directly related to the dynamic states of the system defined by Eqs. (7.29)–(7.40).

Specific growth rates are defined by Monod equations. The phase variables and equation parameters are given in Table 7.4 and their values are defined in [13].

7.3.2 Neural Identifier

Artificial neural networks have been used successfully in biological processes for nonlinear dynamics modeling, prediction, and control [35,25]. RHONNs have proven to be feasible in identification and control applications due to their flexible architecture and robustness [39]. There are different structures of nonlinear neural models that depend on the regression vector and the internal architecture of the network. In this chapter, the NNARX structure is proposed, where the network output vector is defined as the regression vector of an autoregressive external input linear model structure (ARX). The structure trained with past values of the output allows to find

a stable prediction of real-value time series. Nonlinear control systems can be represented as follows:

$$y_{k+1} = F(x_k) + Gu_k, \tag{7.43}$$
$$F(x_k) = F(y_k, y_{k-1}, \ldots, y_{k-z+1}), \tag{7.44}$$

where y_k is the output of the system, F is a nonlinear function, x_k are the states, G is the input matrix, u_k are the inputs to the system, and z is the dimension of the system state space. In this way, the present value of the output y_{k+1} is defined in terms of its past values and the present values of the input. The neural predictor is defined as follows:

$$\hat{y}_{k+1} = \psi(\zeta_k, w) + w'u_k, \tag{7.45}$$
$$\zeta_k = [y_k, \ldots, y_{k-z+1}]^\tau, \tag{7.46}$$

where $\hat{y}_{k+1}$ is the estimated output, ψ is a nonlinear function, ζ_k is the output regression vector, w is the vector of the adjustable weights of the network, and w' is a fixed weight vector for inputs. The neural network is trained with the EKF algorithm to update weights during training [39]. For EKF neural network training, the network weights become the states to be estimated. In this case the error between the neural network output and the measured plant output can be considered as additive white noise. The goal is to find the optimal weights that minimize the prediction error. The weight vector is constructed with the weights of the hidden layer and the output layer. Fig. 7.5 shows the neural identification scheme for the anaerobic digestion reduced model.

The main advantages for using RHONN in system identification are that (a) we can obtain a model with physical meaning, which is not possible with feedforward neural networks; (b) in the neural model, the state of each neuron represents a state variable of the system to be modeled, which means a considerable reduction in the number of neurons in the network [39]; (c) the

TABLE 7.5 Values of D_1 and D_2.					
Input	Unit	0–500 h	500–1000 h	1000–1500 h	1500–2000 h
D_1	h^{-1}	0.005	0.01	0.015	0.02
D_2	h^{-1}	0.00113	0.00226	0.00339	0.00452

structure trained with past values of the output allows to find a stable prediction of real-value time series, and (d) the neural network structure affinity allows for optimal control design.

7.3.3 NNARX Structures

The neural structure for the reduced anaerobic digestion model is NNARX, where the desired outputs are the gas flows and the inputs are the dilution rates and the substrates. The microorganisms and substrate complex dynamics in the process and directly related to the biofuel production are identified. The state-space anaerobic digestion neural model is represented as follows:

$$\begin{aligned}
x_{1,k+1} &= f_1(x_k), \\
x_{2,k+1} &= f_2(x_k) + g(x_k)u(x_k), \\
y_k &= h(x_k),
\end{aligned} \qquad (7.47)$$

where $x \in R^n$ is the estimated state of the system at time $k \in N$ (N denotes the set of nonnegative integers), $u, y \in R^m$ are the input and the output system, respectively, and $f : R^n \to R^n$, $g : R^n \to R^{n \times m}$, $h : R^n \to R^m$, and $J : R^n \to R^{m \times m}$ are smooth and bounded mappings. We assume $f(0) = 0$ and $h(0) = 0$. The neural network structure is composed of five neurons in the hidden layer with a sigmoidal activation function and one neuron in the output layer with a linear activation function. This structure is selected based on the analytic study of the process. The specific target prediction error is 10^{-4}. The neural network training is performed with an EKF in a series-parallel array, the sampling time is one hour, and the total number of samples is 2000.

The identification is subjected to incremental external disturbances in the inputs and over a period of 2000 hours, as shown in Table 7.5.

Additionally, in order to test the process sensibility in the presence of input disturbances, the time-varying input concentration of composites S_{5in} is introduced. Figs. 7.15–7.22 display the identification results.

As can be seen, the proposed neural network efficiently identifies the complex dynamics of the anaerobic digestion system. The s_1, s_2, s_3, x_1, and x_2 dynamics correspond to the substrates and microorganisms

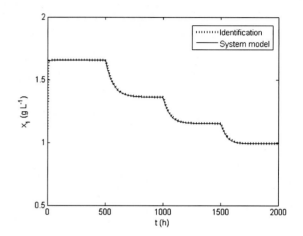

FIG. 7.15   Identification of acidogenic microorganisms.

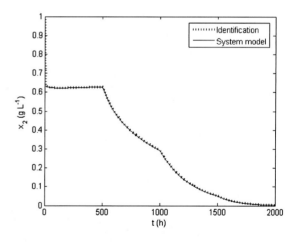

FIG. 7.16   Identification of acetogenic microorganisms.

involved in the hydrogen production in the first reactor. The s_9 and x_3 dynamics correspond to the substrates and microorganisms involved in the production of methane in the second reactor. The hydrogen and methane production is presented in Fig. 7.22.

The mean square error (MSE) reached in the identification is presented in Table 7.6. The results show

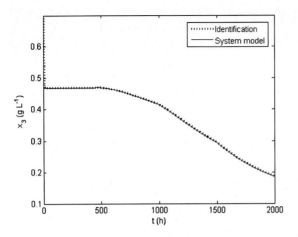

FIG. 7.17   Identification of methanogenic microorganisms.

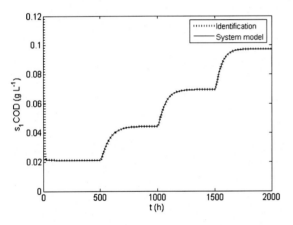

FIG. 7.18   Identification of monosaccharide substrates.

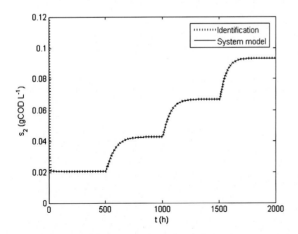

FIG. 7.19   Identification of amino acid substrates.

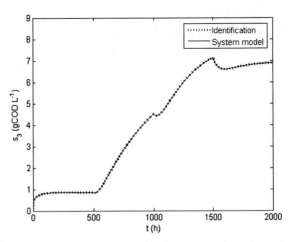

FIG. 7.20   Identification of fatty acid substrates.

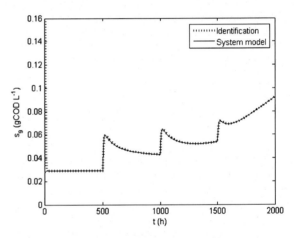

FIG. 7.21   Identification of total acetate.

that the recurrent neural network training with the EKF efficiently captures the complexity associated with the anaerobic process dynamics.

7.3.4 Input–Output Stability Analysis Via Simulation

For the output response analysis of the system, D_1 and D_2 inputs are selected as step, ramp, and sine. The operation range for the inputs is selected from the equilibrium point analysis for the maximum biofuel production [13].

7.3.4.1 Step Input With Random Noise

In this test and from now on, the state response of the system is omitted and only biofuel production is presented, as this is the main objective of this study. Output

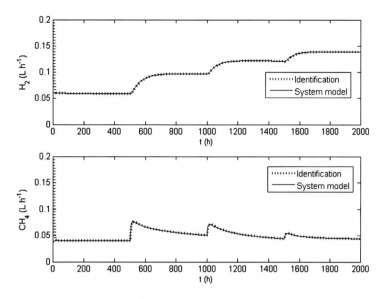

FIG. 7.22 Identification of H_2 and CH_4 production.

TABLE 7.6 Identification MSE.			
State	**MSE**	**State**	**MSE**
x_1	3.1244×10^{-04}	s_2	4.8785×10^{-06}
x_2	3.2571×10^{-05}	s_3	4.5908×10^{-04}
x_3	9.2864×10^{-05}	s_9	6.4557×10^{-05}
s_1	2.8500×10^{-06}		

response of the biofuel production under step inputs with random noise is presented in Fig. 7.23 and the input values are presented in Table 7.7.

As can be seen, the neural network presents a good performance to capture the dynamic behavior of the biofuels in the presence of random noise. The biofuel production reaches its steady-state values close to 2000 h with a small transient state.

7.3.4.2 Sine Input

In this test, the neural model output response is analyzed with sine input; the parameter values are presented in Table 7.8. The output response of the system is presented in Fig. 7.24.

In this study, hydrogen production and methane production present transient states during the simulation due to the amplitude and frequency of the sine inputs. The neural network properly identifies the dynamic behavior of the biofuel.

7.3.4.3 Ramp Input

In this test, the dynamic response of the biofuel production is analyzed under ramp inputs. Fig. 7.25 displays the estimated hydrogen and methane production trajectories and the input values are presented in Table 7.9 (α and β are positive constants).

The neural network identifies the dynamic behavior of the biofuels with an incremental production along simulation time. A transient state zone is presented up to 1200 h for both biofuels, afterwards, a small increase in the H_2 production and a small decrease in the CH_4 production is observed. The neural identifier is suitable for this test with neglected estimation error. This is due to stable equilibrium points of the system for admissible operation interval.

7.3.5 Stability Analysis in the Presence of Disturbances

7.3.5.1 Ramp Input Disturbance

The next analysis presents the system dynamics when D_1 and D_2 are fixed as constant and S_{5in} is a ramp input disturbance as presented in Table 7.10 (γ is a positive constant).

The neural identification of hydrogen and methane production is presented in Fig. 7.26. In this test, hydrogen production shows an increase as compared with Fig. 7.25. Methane production shows a slow increase since the beginning. As before, this is due to stable equilibrium points of the system for admissible operation interval. The neural network identifies the complex dy-

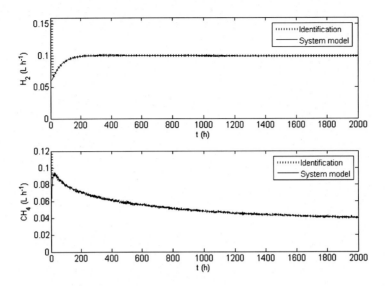

FIG. 7.23 H_2 and CH_4 estimation with step input.

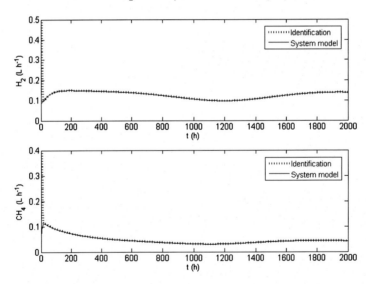

FIG. 7.24 H_2 and CH_4 estimation with sine input.

TABLE 7.7
D_1 and D_2 values.

Input	Unit	Value
D_1	h^{-1}	0.01
D_2	h^{-1}	0.00226

TABLE 7.8
D_1 and D_2 values.

Input	Unit	Value
D_1	h^{-1}	$A * sin(\omega * t)$ $A = 0.00113, \quad \omega = 0.0042$
D_2	h^{-1}	$A * sin(\omega * t)$ $A = 0.005, \quad \omega = 0.0042$

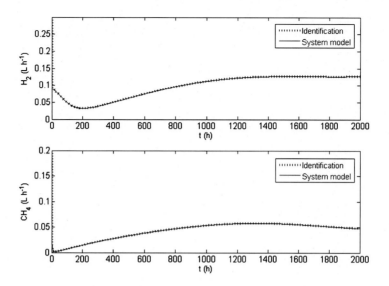

FIG. 7.25 H_2 and CH_4 estimation with ramp input.

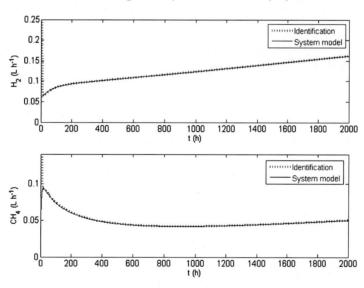

FIG. 7.26 Biofuel estimation in the presence of ramp input disturbance.

TABLE 7.9
D_1 and D_2 values.

Input	Unit	Value
D_1	h^{-1}	(αt)
D_2	h^{-1}	(βt)

TABLE 7.10
D_1 and D_2 values.

Input	Unit	Value
D_1	h^{-1}	0.0125
D_2	h^{-1}	0.002825
S_{5in}	g COD L^{-1}	$(\gamma t) + 40$

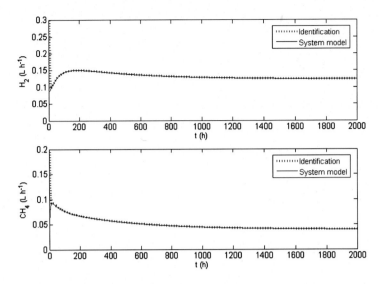

FIG. 7.27 Biofuel estimation in the presence of step input disturbance.

TABLE 7.11
D_1 and D_2 values.

Input	Unit	Value
D_1	h^{-1}	0.0125
D_2	h^{-1}	0.002825
S_{5in}	g COD L^{-1}	56.25

namics in the presence of disturbances with a neglected transient error during the simulation.

7.3.5.2 Step Input Disturbance

The second test consists in a step input disturbance for S_{5in} when D_1 and D_2 are fixed as constants. The parameter values are presented in Table 7.11 and the simulation results are displayed in Fig. 7.27.

In this test, biofuels production presents a transient state zone up to 1200 h and afterwards a stable production zone is reached. The artificial neural network properly identifies the biofuel production in the presence of step disturbances with a neglected transient state error.

7.4 CONCLUSION

In this work, high-order neural network structures for unknown nonlinear systems in the presence of external disturbances and parameter uncertainties for complex dynamics identification are proposed. Two biological models, derived from the ADM1 and ASM1 basic structure, respectively, where assumptions were made to obtain reduced models, maintaining the relevant phenomena of the processes, are proposed. The biological processes are highly nonlinear and submitted to external disturbances and parameter uncertainties, which affect stability and optimal yields. The objective is modeling and predicting dynamic variables that are hardly measurable and relatively expensive or time consuming, but which are necessary for process supervision and control. In the first model, a recurrent high-order neural observer structure is proposed to estimate the complex states of the aerobic digestion process related to the effluent quality at the output. An analysis of the organic compounds reduction at the system output in the presence of disturbances is presented. In the second model, a recurrent neural network with autoregressive external input structure is proposed to predict the complex anaerobic digestion states related to biofuel production. An analysis of the gaseous streams at the system output in the presence of disturbances is presented. In both cases, the neural network structure is trained with an EKF algorithm and the estimated variables are calculated based on the measurable output of the system. The proposed recurrent neural network structures efficiently identify complex dynamics of the process in the presence of external disturbances, which are hard to measure in continuous time. The neural mathematical models have an affine form, which permits the applicability of advanced control techniques.

REFERENCES

1. E. Aguilar-Garnica, J.P. García-Sandoval, D. Dochain, Monitoring of a biodiesel production process via reset observer, Journal of Process Control 42 (2016) 104–113.
2. A.Y. Alanis, M. Hernandez-Gonzalez, E.A. Hernandez-Vargas, Observers for biological systems, Applied Soft Computing 24 (2014) 1175–1182.
3. A.Y. Alanis, E.N. Sanchez, Chapter 3 – full order neural observers, in: A.Y. Alanis, E.N. Sanchez (Eds.), Discrete-Time Neural Observers, Academic Press, 2017, pp. 23–74.
4. A.Y. Alanis, E.N. Sanchez, A.G. Loukianov, E.A. Hernandez, Discrete-time recurrent high order neural networks for nonlinear identification, Journal of the Franklin Institute 347 (7) (2010) 1253–1265.
5. A.Y. Alanis, E.N. Sanchez, A.G. Loukianov, M.A. Perez, Real-time recurrent neural state estimation, IEEE Transactions on Neural Networks 22 (3) (2011) 497–505.
6. G. Antonopoulou, H.N. Gavala, I.V. Skiadas, G. Lyberatos, ADM1-based modeling of methane production from acidified sweet sorghum extract in a two stage process, Bioresource Technology 106 (1) (2012) 10–19.
7. M. Araneda, J. Pavez, B. Luza, D. Jeison, Use of activated sludge biomass as an agent for advanced primary separation, Journal of Environmental Management 192 (2017) 156–162.
8. M.K. Arantes, H.J. Alves, R. Sequinel, E.A. da Silva, Treatment of brewery wastewater and its use for biological production of methane and hydrogen, International Journal of Hydrogen Energy 42 (42) (2017) 26243–26256.
9. P. Arroyo, M. Molinos-Senante, Selecting appropriate wastewater treatment technologies using a choosing-by-advantages approach, Science of the Total Environment 625 (2018) 819–827.
10. J. Bai, H. Liu, B. Yin, H. Ma, X. Chen, Modified ADM1 for modeling free ammonia inhibition in anaerobic acidogenic fermentation with high-solid sludge, Journal of Environmental Sciences (China) 52 (2017) 58–65.
11. T. Beltramo, C. Ranzan, J. Hinrichs, Artificial neural network prediction of the biogas flow rate optimised with an ant colony algorithm, Biosystems Engineering 143 (2016) 68–78.
12. F. Benazzi, R. Katebi, Nonlinear observability of activated sludge process models, IFAC 38 (1) (2005).
13. M. Borisov, N. Dimitrova, I. Simeonov, Mathematical modelling of anaerobic digestion with hydrogen and methane production, IFAC-PapersOnLine 49 (26) (2016) 231–238.
14. W. Cao, X. Wang, Z. Ming, J. Gao, A review on neural networks with random weights, Neurocomputing 275 (2018) 278–287.
15. K.K. Castillo-Villar, S. Eksioglu, M. Taherkhorsandi, Integrating biomass quality variability in stochastic supply chain modeling and optimization for large-scale biofuel production, Journal of Cleaner Production 149 (2017) 904–918.
16. G. De Gioannis, A. Muntoni, A. Polettini, R. Pomi, D. Spiga, Energy recovery from one- and two-stage anaerobic digestion of food waste, Waste Management 68 (2017) 595–602.
17. J. Deng, Dynamic neural networks with hybrid structures for nonlinear system identification, Engineering Applications of Artificial Intelligence 26 (1) (2013) 281–292.
18. A.K. Dhussa, S.S. Sambi, S. Kumar, S. Kumar, S. Kumar, Nonlinear Autoregressive Exogenous modeling of a large anaerobic digester producing biogas from cattle waste, Bioresource Technology 170 (2014) 342–349.
19. S. Diop, E. Chorukova, I. Simeonov, Modeling and specific growth rates estimation of a two-stage anaerobic digestion process for hydrogen and methane production, IFAC-PapersOnLine 50 (1) (2017) 12641–12646.
20. Z. Duan, M.N. Cruz Bournazou, C. Kravaris, Dynamic model reduction for two-stage anaerobic digestion processes, Chemical Engineering Journal 327 (2017) 1102–1116.
21. A. Ersöz, Y. DurakÇetin, A. Sarıoğlan, A.Z. Turan, M.S. Mert, F. Yüksel, H.E. Figen, N. Güldal, M. Karaismailoğlu, S.Z. Baykara, Investigation of a novel & integrated simulation model for hydrogen production from lignocellulosic biomass, International Journal of Hydrogen Energy 43 (2) (2018) 1081–1093.
22. C. Foscoliano, S. Del, M. Mulas, S. Tronci, Predictive control of an activated sludge process for long term operation, Chemical Engineering Journal 304 (2016) 1031–1044.
23. S. Frigo, G. Spazzafumo, Cogeneration of power and substitute of natural gas using biomass and electrolytic hydrogen, International Journal of Hydrogen Energy 43 (26) (2018) 11696–11705.
24. S.F. Fu, X.H. Xu, M. Dai, X.Z. Yuan, R.B. Guo, Hydrogen and methane production from vinasse using two-stage anaerobic digestion, Process Safety and Environmental Protection 107 (189) (2017) 81–86.
25. K.J. Gurubel, V. Osuna-Enciso, A. Coronado-Mendoza, E. Cuevas, Optimal control strategy based on neural model of nonlinear systems and evolutionary algorithms for renewable energy production as applied to biofuel generation, Journal of Renewable and Sustainable Energy 9 (3) (2017).
26. C. Gustavsson, C. Hulteberg, Co-production of gasification based biofuels in existing combined heat and power plants – analysis of production capacity and integration potential, Energy 111 (2016) 830–840.
27. S. Hassam, E. Ficara, A. Leva, J. Harmand, A generic and systematic procedure to derive a simplified model from the anaerobic digestion model No. 1 (ADM1), Biochemical Engineering Journal 99 (2015) 193–203.
28. S. Hwangbo, S.K. Heo, C. Yoo, Network modeling of future hydrogen production by combining conventional steam methane reforming and a cascade of waste biogas treatment processes under uncertain demand conditions, Energy Conversion and Management 165 (2018) 316–333.
29. J. Mohd, N.H. Hoang, M.A. Hussain, D. Dochain, Review and classification of recent observers applied in chemical process systems, Computers & Chemical Engineering 76 (2015) 27–41.
30. J. Mohd, N.H. Hoang, M.A. Hussain, D. Dochain, Hybrid observer for parameters estimation in ethylene polymerization reactor: a simulation study, Applied Soft Computing 49 (2016) 687–698.

31. J. Mohd, M.A. Hussain, M.O. Tade, J. Zhang, Artificial Intelligence techniques applied as estimator in chemical process systems – a literature survey, Expert Systems with Applications 42 (14) (2015) 5915–5931.

32. V. Montiel Corona, E. Razo-Flores, Continuous hydrogen and methane production from Agave tequilana bagasse hydrolysate by sequential process to maximize energy recovery efficiency, Bioresource Technology 249 (2018) 334–341.

33. M.I. Nelson, H.S. Sidhu, Analysis of the activated sludge model (number 1), Applied Mathematics Letters 22 (5) (2009) 629–635.

34. M.I. Oloko-Oba, A.E. Taiwo, S.O. Ajala, B.O. Solomon, E. Betiku, Performance evaluation of three different-shaped bio-digesters for biogas production and optimization by artificial neural network integrated with genetic algorithm, Sustainable Energy Technologies and Assessments 26 (2018) 116–124.

35. J.D. Rios, A.Y. Alanis, C. Lopez-Franco, N. Arana-Daniel, RHONN identifier-control scheme for nonlinear discrete-time systems with unknown time-delays, Journal of the Franklin Institute 355 (1) (2018) 218–249.

36. A. Rodríguez, G. Quiroz, R. Femat, H.O. Méndez-Acosta, J.D. León, An adaptive observer for operation monitoring of anaerobic digestion wastewater treatment 269 (2015) 186–193.

37. A. Román-Martínez, P. Lanuza-Pérez, M. Cepeda-Rodríguez, Global sensitivity analysis for control of biological wastewater treatment 33 (2014), Figure 1.

38. M. Salgot, M. Folch, Wastewater treatment and water reuse, Current Opinion in Environmental Science & Health (2018).

39. E.N. Sanchez, A.Y. Alanis, A.G. Loukianov, Discrete-Time High Order Neural Control Trained with Kalman Filtering, Springer-Verlag, Berlin, 2008.

40. I. Satar, W.R.W. Daud, B.H. Kim, M.R. Somalu, M. Ghasemi, Immobilized mixed-culture reactor (IMcR) for hydrogen and methane production from glucose, Energy 139 (2017) 1188–1196.

41. SEMARNAT, Proy-NOM-001-SEMARNAT-2017, 2018, p. 17.

42. F.M. Silva, C.F. Mahler, L.B. Oliveira, J.P. Bassin, Hydrogen and methane production in a two-stage anaerobic digestion system by co-digestion of food waste, sewage sludge and glycerol, Waste Management (2018).

43. I.Y. Smets, J.V. Haegebaert, R. Carrette, J.F.V. Impe, Linearization of the activated sludge model ASM1 for fast and reliable predictions 37 (2003) 1831–1851.

44. J. Sun, B.J. Ni, K.R. Sharma, Q. Wang, S. Hu, Z. Yuan, Modelling the long-term effect of wastewater compositions on maximum sulfide and methane production rates of sewer biofilm, Water Research 129 (2018) 58–65.

45. Q. Wang, W. Wei, Y. Gong, Q. Yu, Q. Li, J. Sun, Z. Yuan, Science of the total environment technologies for reducing sludge production in wastewater treatment plants: state of the art, Science of the Total Environment 587–588 (2017) 510–521.

46. M. Winter, C. Breitsamter, Nonlinear identification via connected neural networks for unsteady aerodynamic analysis, Aerospace Science and Technology 1 (2018) 1–17.

47. W. Yu, Nonlinear system identification using discrete-time recurrent neural networks with stable learning algorithms 158 (2004) 131–147.

48. Y. Zhang, S. Piccard, W. Zhou, Improved ADM1 model for anaerobic digestion process considering physico-chemical reactions, Bioresource Technology 196 (2015) 279–289.

CHAPTER 8

Learning-Based Identification of Viral Infection Dynamics

GUSTAVO HERNANDEZ-MEJIA, PHD • ESTEBAN A. HERNANDEZ-VARGAS, PHD

CHAPTER POINTS

- Theoretical bases of neural networks, such as the structure and training algorithm based on the extended Kalman filter, are introduced.
- The problems of influenza and HIV infections are described, including mathematical models at the within-host level and their characteristics.
- High-order neural networks are used to identify the influenza dynamics and the HIV infection, respectively.
- Identification results describe the performance to different infections.

8.1 INTRODUCTION

Throughout history, high death tolls associated with infectious diseases have appeared around the world. One of the disasters in human history was caused by a viral infection, the 1918 flu pandemic. The result of this catastrophe was approximately 20 million dead people, showing that respiratory tract infections by viral pathogens are one of the biggest health threats to humanity [1]. In particular, influenza A virus (IAV) outbreaks have caused high rates of morbidity and mortality worldwide, leading to critical public health problems. Annual epidemics are estimated to result from 3 to 5 million cases of severe illness and about 500,000 deaths, where pregnant women, toddlers, and seniors are the principal affected groups with serious complications [2]. The Food and Drug Administration (FDA) has approved neuraminidase inhibitors such as oseltamivir and zanamivir for IAV treatment [3,4]. The World Health Organization (WHO) suggests IAV treatment with two levels of medication, curative and pandemic. Curative medication is with fixed doses of 75 mg twice a day, the pandemic level is with 150 mg twice per day, both using oseltamivir. The medication can start within the first three days after infection [3,5].

Another instance of the consequences of the threats represented by emerging viral diseases is HIV/AIDS. The burden of this viral infection has resulted in more than 35 million infected from all socio-economic backgrounds [6]. A common HIV infection response consists of three phases: an initial acute infection, a long asymptomatic period, and a final increase in viral load with the simultaneous collapse in healthy CD4+ T cell counts [7–9]. The combined antiretroviral treatment (cART) suppresses the virus to undetectable levels in the blood, improves immunologic parameters, minimizes the emergence of resistance, and protects from other infections [10].

The design of model-based strategies to control infected host diseases employing identification tools may bring faster results for the medical community. However, the development of mathematical models of infectious diseases is an arduous and complex assignment. Therefore, it is important to incorporate new techniques that allow the identification of viral infections and immune system dynamics. Theoretical and applied identification processes based on recurrent high-order neural networks (RHONNs) have been broadly studied, showing a good performance [11–13]. Also, algorithms based on the extended Kalman filter (EKF) have been introduced to neural networks training [11,14]. Working with foundations of the EKF algorithm, the learning convergence can be augmented for feedforward and recurrent neural networks [13,15]. Neural networks for influenza infection have been proposed for diverse applications, such as transmission between populations, vaccination, and mutation [16–19]. On the basis of RHONNs proposed by [11], here we identify the within-host dynamics of both influenza and HIV infections. This chapter is mainly based on our previous work [20]. The RHONN is used to identify the mathematical model which is considered unknown and assuming that the complete state variables are available. This framework may assist to pave the way for clinical decision making.

8.2 NEURAL IDENTIFICATION

The identification problem is described as follows for a general form of discrete-time nonlinear systems:

$$x(k+1) = F(x(k), u(k)), \qquad (8.1)$$

where $x(k) \in \Re^n$ is the state vector of the plant, $u(k) \in \Re^m$ is the control input, and $F : \Re^n \times \Re^m \to \Re^n$ is a nonlinear function. To identify the discrete-time nonlinear system (8.1), we used the following series-parallel discrete-time RHONN structure [21]:

$$\chi_i(k+1) = \omega_i^T z_i(x(k), u(k)) + \epsilon_{zi}, \qquad (8.2)$$

where $i = 1, \ldots, n$, χ_i is the state of the i-th neuron. ω_i is the on-line adapted weight vector. ϵ_{zi} is a bounded approximation error and $z_i(x(k), u(k))$ is given by

$$z_i(x(k), u(k)) = \begin{pmatrix} z_{i_1} \\ z_{i_2} \\ \vdots \\ z_{i_{L_i}} \end{pmatrix} = \begin{pmatrix} \prod_{j \in I_1} \xi_{i_j}^{d_{i_j}^{(1)}} \\ \prod_{j \in I_2} \xi_{i_j}^{d_{i_j}^{(2)}} \\ \vdots \\ \prod_{j \in I_{L_i}} \xi_{i_j}^{d_{i_j}^{(L_i)}} \end{pmatrix}, \qquad (8.3)$$

$$\xi_i = \begin{pmatrix} \xi_{i_1} \\ \vdots \\ \xi_{i_n} \\ \xi_{i_{n+1}} \\ \vdots \\ \xi_{i_{n+m}} \end{pmatrix} = \begin{pmatrix} \varphi(x_1) \\ \vdots \\ \varphi(x_n) \\ u_1 \\ \vdots \\ u_m \end{pmatrix}, \qquad (8.4)$$

$$\varphi(v) = \frac{1}{1 + e^{-av}}, \quad a > 0. \qquad (8.5)$$

L_i is the respective number of high-order connections, $\{ I_1, I_2, \ldots, I_{L_i} \}$ is a collection of nonordered subsets of $\{ 1, 2, \ldots, n+m \}$, with $d_{ij}(k)$ being a nonnegative integer. $u = \{u_1, u_2, \ldots u_m\}^T$ in (8.4) is the input vector to the neural network with $\varphi(\cdot)$ defined by (8.5), where v can be defined as a real-value variable [15]. The RHONN (8.2) considers ϵ_{zi} a bounded approximation error, which can be reduced by increasing the number of the adjustable weights [21]. Assume the existence of an ideal

weight vector ω_i^* such that $\| \epsilon_{zi} \|$ can be minimized on a compact set $\Omega_{z_i} \subset \Re^{L_i}$. The ideal weight vector ω_i^* is an artificial quantity for analytical approaches and it is assumed that this vector exists and is constant but unknown. Additional details can be found in [11,13,21].

8.2.1 Network Training Based on the Extended Kalman Filter

The training is performed on-line by the modified EKF-based algorithm [22], as follows:

$$\begin{align} \omega_i(k+1) &= \omega_i(k) + \eta_i K_i(k) e_i(k), & (8.6) \\ K_i(k) &= P_i(k) H_i(k) M_i(k), & (8.7) \\ P_i(k+1) &= P_i(k) - K_i(k) H_i^T(k) P_i(k) + Q_i(k), & (8.8) \end{align}$$

with

$$\begin{align} M_i(k) &= [R_i(k) + H_i^T(k) P_i(k) H_i(k)]^{-1}, & (8.9) \\ e_i(k) &= x_i(k) - \chi_i(k), & (8.10) \\ H_{ij} &= \left[\frac{\partial \chi_i(k)}{\partial \omega_{ij}(k)} \right]^T, & (8.11) \end{align}$$

where $i = 1, \ldots, n$ and n is the number of states. $e_i(k) \in \Re$ is the respective identification error. $P_i(k) \in \Re^{L_i \times L_i}$ is the prediction error–associated covariance matrix at the step k. $\omega_i \in \Re^{L_i}$ is the weight vector considered as state of the network, L_i is the respective number of neural network weights. χ_i is the i-th neural network state, x_i is the i-th plant state. $K_i \in \Re^{L_i}$ is the Kalman gain vector. $Q_i \in \Re^{L_i \times L_i}$ is the state noise–associated covariance matrix. $R_i \in \Re$ is the measurement noise–associated covariance. $H_i \in \Re^{L_i}$ is a vector, in which each entry H_{ij} is the derivative of one of the neural network states χ_i, with respect to one neural network weight ω_{ij} defined in (8.11), with $i = 1, \ldots, n$ and $j = 1, \ldots, L_i$. A common practice is that P_i and Q_i are initialized as diagonal matrices, with entries $P_i(0)$ and $Q_i(0)$, respectively [13,15]. It is important to remark that $H_{ij}(k)$, $P_i(k)$, and $K_i(k)$ for the EKF are bounded [22].

Fig. 8.1 shows the identification scheme where $u(k)$ represents the drug intake, if necessary, to the within-host model as well as to the RHONN. The identification error signal $e(k)$ is the criterion for EKF training as in (8.10), formed with the difference between sampled within-host model states $x(k)$ and RHONN identified states $\chi(k)$. Finally, neural weights $w(k)$ are adapted on-line.

8.3 WITHIN-HOST INFLUENZA INFECTION

In this chapter, we employ the IAV model proposed in [23]. This model consists of the CD8+ T cell (E) and

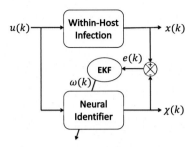

FIG. 8.1 RHONN identification scheme for a within-host infection model.

virus (V) dynamics at a host level. Additionally, we incorporate the drug dynamics (D) covering two drug phases, such as pharmacokinetics (PK) and pharmacodynamics (PD). This provides a framework for linking interactions between drugs and pharmacological targets [24,25]. The influenza model is formed with the following differential equations:

$$\dot{E} = S_E + rE\left(\frac{V}{V + k_e}\right) - c_e E, \quad (8.12)$$

$$\dot{V} = pV\left(1 - \frac{V}{k_v}\right) - c_v V E. \quad (8.13)$$

The parameter k_v is the maximum carrying capacity of the virus. p is the virus replication rate. The virus is cleared with a rate $c_v E$. The CD8+ T cells constant replenishment rate is $S_E = c_e E(0)$, where c_e is the half-life of CD8+ T cells and $E(0)$ is the initial number of CD8+ T cells. r is the CD8+ T cell replication rate. k_e is the half saturation constant of E cell proliferation. The parameters values are the following [23]: r (0.33 day^{-1}), k_e (2.7 × 10^3 PFU/mL), c_e (2.0 × 10^{-2} day^{-1}), p (4.4 day^{-1}), k_v (10^6 PFU/mL), and c_v (1.24 × 10^{-6} day^{-1}cell^{-1}). The initial state values are $E(0)$ (10^6 cells) and $V(0)$ (25 PFU/mL) [23].

For drug dynamics, the temporal distribution of drug concentration in different organs is described by the PK phase. This is modeled by a constant decay differential equation, i.e.,

$$\dot{D} = -\delta_D D, \quad (8.14)$$

where D is the amount of drug available and δ_D is the drug elimination rate. Additionally, the PD phase describes the effect of a drug on the organism. PD dynamics can be represented by

$$\eta = \frac{D}{D + EC_{50}}, \quad (8.15)$$

indicating the efficacy of the treatment, where the parameter EC_{50} is the drug concentration level that provides 50% of the drug efficacy. The drug intake time is defined by τ_k, indicating the sequence of drug intakes. To link the antiviral therapy with the influenza model, we consider that the viral replication rate p can be replaced by $p(1 - \eta)$. This is based on the fact that antiviral therapies mainly affect the replication cycle of the virus. Consequently, Eq. (8.13) can be rewritten as

$$\dot{V} = p\left(1 - \frac{D}{D + EC_{50}}\right)V\left(1 - \frac{V}{k_v}\right) - c_v V E, \quad (8.16)$$

with

$$\dot{D}(t) = -\delta_D D(t), \quad \tau_k \leq t < \tau_{k+1}, \quad (8.17)$$

with $\tau_k \leq t < \tau_{k+1}$ as the time intervals between drug intakes. This indicates that the first drug intake of the treatment is given by $D(\tau_1) = D(t_0)$. PK/PD parameter values such as EC_{50} (42.30 mg) and δ_D (3.26 day^{-1}) are reported in [27].

An on-line neural model is developed for identifying IAV dynamics including the virus, CD8+ T cells, and drug dynamics. The proposed neural structure is

$$\chi_E(k+1) = \omega_{11}(k)\varphi(x_E(k)) + \omega_{12}(k)\varphi(x_V(k)) + g_1 u(k), \quad (8.18)$$

$$\chi_V(k+1) = \omega_{21}(k)\varphi(x_E(k)) + \omega_{22}(k)\varphi(x_V(k)) + \omega_{23}(k)\varphi(x_D(k)) + g_2 u(k), \quad (8.19)$$

$$\chi_D(k+1) = \omega_{31}(k)\varphi(x_D(k)) + g_3 u(k), \quad (8.20)$$

where $\varphi(x_i)$ has the form of (8.5). Neural weights ω_{ij} are adapted on-line with the EKF-based algorithm in (8.6), (8.7), and (8.8). Also, g_1, g_2, and g_3 are adjustable gains for RHONN input $u(k)$, which in this case is the drug. In addition, as the neural network structure is in discrete time, variables $x_E(k)$, $x_V(k)$, and $x_D(k)$ represent the sampled states from the IAV model. Note that the CD8+ T cell state (E) in (8.12) is represented by $\chi_E(k)$ in (8.18), the influenza virus state (V) in (8.16) is related to $\chi_V(k)$ in (8.19), and the drug (D) in (8.17) is presented by $\chi_D(k)$ in (8.20).

8.4 WITHIN-HOST HIV INFECTION

The model proposed by [8] is herein employed to develop an HIV-RHONN identifier. This model describes the dynamics of the healthy (uninfected) CD4+ T cells (T), the infected active CD4+ T cells (T^*), the uninfected macrophages (M), the infected macrophages

(M^*), and the HIV (V_H). The model is written as follows:

$$\dot{T} = s_T + \frac{\rho_T V_H}{C_T + V_H}T - k_T T V_H - \delta_T T, \quad (8.21)$$

$$\dot{T}^* = k_T T V_H - \delta_{T^*} T^*, \quad (8.22)$$

$$\dot{M} = s_M + \frac{\rho_M V_H}{C_M + V_H}M - k_M M V_H - \delta_M M, \quad (8.23)$$

$$\dot{M}^* = k_M M V_H - \delta_{M^*} M^*, \quad (8.24)$$

$$\dot{V}_H = p_T T^* + p_M M^* - \delta_V V_H, \quad (8.25)$$

where ρ_T (0.01 day^{-1}) and ρ_M (0.003 day^{-1}) are the growth rate of CD4+ T cells and macrophages with a carrying capacity C_T (300 copies/mm^3) and C_M (220 copies/mm^3), respectively. k_T (4.57 × 10^{-5} mm^3/day copies) is the rate at which free viruses V_H infect CD4+ T cells. δ_T (0.01 day^{-1}) stands for the death rate of CD4+ T cells. s_T (10 cells/mm^{3}day) and s_M (0.15 cells/mm^{3}day) are the new cell source terms. δ_{T^*} (0.4 day^{-1}) is the clearance of infected CD4+ T cells. The rate at which free viruses V_H infect macrophages is k_M (4.33 × 10^{-8} mm^3/day copies). δ_M (1 × 10^{-3} day^{-1}) and δ_{M^*} (1 × 10^{-3} day^{-1}) stand for the death rate of uninfected and infected macrophages, respectively. Finally, the amount of virus produced by infected CD4+ T cells and macrophages is given by p_T (38 copies/cell day) and p_M (35 copies/cell day), respectively. δ_V (2.4 day^{-1}) is the clearance rate of viral particles. CD4+ T cells are taken as 1000 cells/mm^3 and 150 cells/mm^3 for macrophages. Infected cells are initially zero and the initial viral concentration are 10 copies/mL [8].

The structure of the HIV-RHONN to identify the model (8.21)–(8.25) is as follows:

$$\chi_T(k+1) = \omega_{11}(k)\varphi(x_T(k)) + \omega_{12}(k)\varphi(x_T(k))\varphi(x_{V_H}(k)), \quad (8.26)$$

$$\chi_{T^*}(k+1) = \omega_{21}(k)\varphi(x_T(k))\varphi(x_{V_H}(k)) + \omega_{22}(k)\varphi(x_{T^*}(k)), \quad (8.27)$$

$$\chi_M(k+1) = \omega_{31}(k)\varphi(x_M(k))\varphi(x_{V_H}(k)) + \omega_{32}(k)\varphi(x_V(k)), \quad (8.28)$$

$$\chi_{M^*}(k+1) = \omega_{41}(k)\varphi(x_{M^*}(k))\varphi(x_{V_H}(k)) + \omega_{42}(k)\varphi(x_{M^*}(k)), \quad (8.29)$$

$$\chi_{V_H}(k+1) = \omega_{51}(k)\varphi(x_{T^*}(k)) + \omega_{52}(k)\varphi(x_{M^*}(k)) + \omega_{53}(k)\varphi(x_{V_H}(k)), \quad (8.30)$$

where $x_T(k)$, $x_{T^*}(k)$, $x_M(k)$, $x_{M^*}(k)$, and $x_{V_H}(k)$ are the sampled states from the HIV model and $\varphi(x_i)$ is defined by (8.5). We opted for a simple RHONN structure,

the reasoning is that each of the identified states, χ_i, is related to the HIV model states. For instance, viral dynamics in (8.25) is dependent on states T^* and M^* and itself, so the neural state χ_{V_H} is a function of the same sampled HIV model states. This fashion is consistent with the rest of the HIV-RHONN equations.

8.5 NUMERICAL RESULTS

Numerical results are presented for both identifiers, the IAV-RHONN, and the HIV-RHONN. Also, we introduce the parameter values for the EKF training algorithm initiation.

8.5.1 IAV-RHONN Identification

The results of the IAV-RHONN performance are plotted in Figs. 8.2 to 8.4, presenting the dynamics of the influenza model and the behavior of the neural network identification. The IAV-RHONN is able to identify the entire dynamics of the IAV model for both treatment levels, curative and pandemic. Additionally, the network allows the identification at different initiation days of medication. We fine-tuned parameters for the EKF for each of the identified states. For instance, the parameters for CD8+ T cells are

$$Q_E = 1 \times 10^6 \begin{bmatrix} 1 & 0 \\ 0 & 1 \end{bmatrix}, \ R_E = 1 \times 10^4, \ \eta_E = 0.93, \quad (8.31)$$

and initial values are

$$P_E = \begin{bmatrix} 10 & 0 \\ 0 & 10 \end{bmatrix}, \ \omega_E = rand(2,1) * 1200, \quad (8.32)$$

where the initial value of ω_E is calculated with the command *rand* from Matlab. For virus dynamics, we have the following values:

$$Q_V = 1 \times 10^6 \begin{bmatrix} 1 & 0 & 0 \\ 0 & 1 & 0 \\ 0 & 0 & 1 \end{bmatrix}, \ R_V = 1 \times 10^4, \ \eta_V = 0.63 \quad (8.33)$$

and the initial values

$$P_V = \begin{bmatrix} 1000 & 0 & 0 \\ 0 & 1000 & 0 \\ 0 & 0 & 1000 \end{bmatrix}, \ \omega_V = rand(3,1) * 0.05. \quad (8.34)$$

Finally, drug parameters for the EKF are

$$Q_D = 1 \times 10^6, \ R_D = 1 \times 10^4, \ \eta_D = 0.13, \quad (8.35)$$

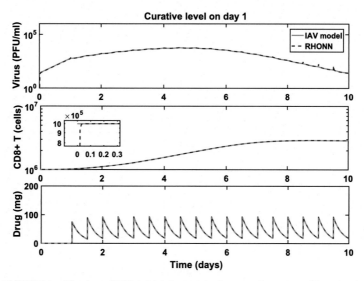

FIG. 8.2 IAV-RHONN identification of IAV model dynamics. Curves of the virus, CD8+ T cells, and drug are presented for starting day 1 with curative FDA treatment. For illustration, a zoom-window shows the initial error of the CD8+ T cells.

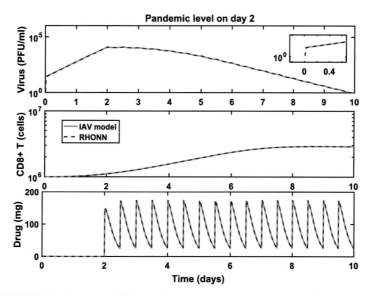

FIG. 8.3 IAV-RHONN identification of IAV model dynamics. Curves of the virus, CD8+ T cells, and drug are presented for the initiation day 2 with pandemic FDA treatment. An error zoom-window of viral dynamics is presented in the viral top panel.

with initial values

$$P_D = 1, \quad \omega_D = rand(1, 1) * 0.05, \qquad (8.36)$$

and $g_1 = g_2 = g_3 = 1$. The sampling time is 0.01 days and the simulation time is 10 days. Fig. 8.2

shows the performance of the IAV-RHONN and the influenza model; in this case a curative treatment level is considered. Dynamics of viral load, CD8+ T cells, and the drug are adequately identified by the IAV-RHONN, which highlights the capacity of the neural network.

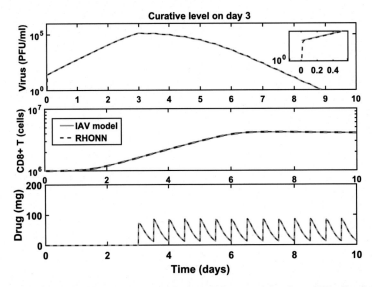

FIG. 8.4 IAV-RHONN identification of IAV model dynamics. Curves of the virus, CD8+ T cells, and drug are presented for the initiation day 3 with curative FDA treatment. A zoom-window shows the initial viral error identification.

Fig. 8.3 presents the dynamics of the IAV model under a pandemic medication level over the second initiation day. We can notice that the virus suddenly changes the behavior with the drug intake on day 2 and the IAV-RHONN is following this change.

The identification of a curative treatment starting on day 3 is presented in Fig. 8.4. Once again, the IAV-RHONN identifies the dynamics of the IAV state variables. At this point, simulations show that the network continues to identify the IAV system after any starting day of treatment and both levels of medication, curative and pandemic. It is important to mention that in practically all simulation tests, the identification error is around 0.01 for the virus, 0.001 for the CD8+ T cells, and up to 1 for the drug. These error values are with the same magnitudes, as shown in Figs. 8.2 to 8.4.

8.5.2 HIV-RHONN Identification

The fine-tuned initial EKF parameters for the uninfected T cells state (T) are as follows:

$$Q_T = 1 \times 10^6 \begin{bmatrix} 1 & 0 \\ 0 & 1 \end{bmatrix}, \ R_T = 1 \times 10^4, \ \eta_T = 1, \ (8.37)$$

and initial values are

$$P_T = \begin{bmatrix} 10 & 0 \\ 0 & 10 \end{bmatrix}, \ \omega_T = rand(2, 1) * 1200. \quad (8.38)$$

For infected T cells dynamics (T^*), we have the followings values:

$$Q_{T*} = 1 \times 10^6 \begin{bmatrix} 1 & 0 & 0 \\ 0 & 1 & 0 \\ 0 & 0 & 1 \end{bmatrix}, \ R_{T*} = 1 \times 10^4, \ \eta_{T*} = 1,$$

$$(8.39)$$

and initial values are

$$P_{T*} = \begin{bmatrix} 10 & 0 & 0 \\ 0 & 10 & 0 \\ 0 & 0 & 10 \end{bmatrix}, \ \omega_{T*} = rand(3, 1). \quad (8.40)$$

The macrophage (M) EKF values are

$$Q_M = 1 \times 10^6 \begin{bmatrix} 1 & 0 \\ 0 & 1 \end{bmatrix}, \ R_M = 1 \times 10^4, \ \eta_M = 1, \quad (8.41)$$

and initial values are

$$P_M = \begin{bmatrix} 10 & 0 \\ 0 & 10 \end{bmatrix}, \ \omega_M = rand(2, 1). \quad (8.42)$$

For infected macrophages (M^*), we set $\eta_{M*} = 1$ and the values

$$Q_{M*} = 1 \times 10^6 \begin{bmatrix} 1 & 0 & 0 \\ 0 & 1 & 0 \\ 0 & 0 & 1 \end{bmatrix}, \ R_{M*} = 1 \times 10^4, \quad (8.43)$$

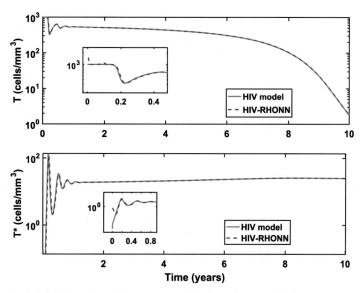

FIG. 8.5 HIV-RHONN identification of CD4+ T cells dynamics. The top panel shows the HIV model and RHONN dynamics of uninfected CD4+ T cells, the bottom panel displays the identification of infected CD4+ T cells. Zoom-windows in both panels stand for the initial identification error.

with initial values

$$P_{M*} = \begin{bmatrix} 10 & 0 & 0 \\ 0 & 10 & 0 \\ 0 & 0 & 10 \end{bmatrix}, \; \omega_{M*} = rand(3, 1) * 0.5.$$

(8.44)

HIV (V_H) identification works with $\eta_{V_H} = 2$ and the parameters

$$Q_{V_H} = 1 \times 10^6 \begin{bmatrix} 1 & 0 & 0 \\ 0 & 1 & 0 \\ 0 & 0 & 1 \end{bmatrix}, \; R_{V_H} = 1 \times 10^4 \quad (8.45)$$

and the initial values

$$P_{V_H} = \begin{bmatrix} 10 & 0 & 0 \\ 0 & 10 & 0 \\ 0 & 0 & 10 \end{bmatrix}, \; \omega_{V_H} = rand(3, 1) * 2. \quad (8.46)$$

The sampling time for the HIV identification is 2 days and the simulation time is 10 years. The neural identification is shown in Figs. 8.5 to 8.7, where each panel of the figures presents a zoom-window highlighting the initial identification error.

For the case of CD4+ T cells, Fig. 8.5 shows the accurate identification of both the uninfected and the infected cell dynamics. It is important to note that, during the first year, although the infected CD4+ T cells dynamics is highly variable, the HIV-RHONN is able to keep acceptable identification performance. For the same period of time, the uninfected cells show minor variations in dynamics, which is also identified by the proposed RHONN.

Fig. 8.6 shows the dynamics of both the infected and the uninfected macrophages. The identification of uninfected macrophages is performed with a larger identification error during the first 0.1 years, compared to that of the infected ones in the same initial period. The zoom-windows depict these errors (top window). It can be observed the good performance of the identifier to represent the original macrophages dynamics. In the bottom window, a negligible identification error is shown for the infected macrophage state.

The HIV dynamics of both the HIV model and the HIV identifier are depicted in Fig. 8.7, where the identification performance can be verified. The initial error is highlighted in the zoom window, which clearly shows the identification during the first samples. The high variability of viral dynamics, especially during the first year of infection, is identified by the proposed HIV-RHONN. Also, after the fifth year, the viral dynamics presents an exponential shape, which is consistently reproduced by the HIV-RHONN model. It is important to mention that in all tests, the identification error is around 0.01 for the HIV-RHONN. This error value is in accordance with the magnitudes shown in Figs. 8.5 to 8.7.

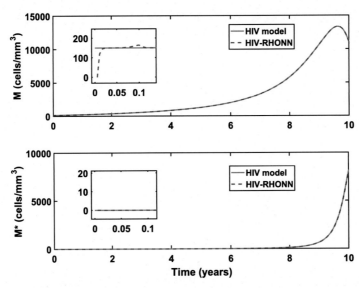

FIG. 8.6 HIV-RHONN identification of macrophage dynamics. The top panel displays the HIV model and RHONN dynamics of the uninfected macrophages, the bottom panel shows the identification of the infected macrophages. Zoom-windows in both panels stand for the initial identification error.

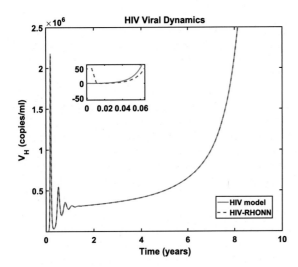

FIG. 8.7 HIV-RHONN identification of viral dynamics. The HIV model and RHONN dynamics are displayed. The zoom-window represents the initial identification error.

8.6 CONCLUSIONS

Two neural identifiers for HIV and influenza infections are trained with the EKF-based algorithm. The neural identification scheme is capable of performing the representation of HIV dynamics with a small identification error. Also, for the influenza infection, diverse tests illustrate the neural network capability, reporting a good

performance for different starting days of treatment and levels of medication. Different infectious diseases may be potentially studied employing the proposed framework [26]. Furthermore, the same identification approach can help to estimate drug PK/PD dynamics in compartmental modeling [27]. Control techniques for drug calculation based on the RHONN dynamics will be explored in the future for personalized treatments [28]. In addition, the observers proposed by [29] may solve estimation issues in clinical procedures.

ACKNOWLEDGMENTS

This work is supported by the Alfons and Gertrud Kassel-Stiftung and the Deutsche Forschungsgemeinschaft (HE7707/5-1).

REFERENCES

1. N. Restifo, Flu: The Story of the Great Influenza Pandemic of 1918 and the Search for the Virus that Caused It, Nature Publishing Group, 2000.
2. WHO, Influenza (Seasonal), Fact sheet N, vol. 211, World Health Organization, 2016.
3. WHO, Guidelines for Pharmacological Management of Pandemic Influenza A(H1N1) 2009 and Other Influenza Viruses, World Health Organization, 2009.
4. W. Taylor, B. Thinh, Oseltamivir is adequately absorbed following nasogastric administration to adult patients with

severe H5N1 influenza, Public Library of Science 3 (3) (2008).

5. FDA, Tamiflu (Oseltamivir Phosphate) Capsules and for Oral Suspension. Detailed View: Safety Labeling Changes Approved by FDA Center for Drug Evaluation and Research, Food and Drug Administration, 2008.

6. W.H.O. Global, HIV Situation, Fact sheet 211, World Health Organization, 2017.

7. A. Fauci, HIV and AIDS: 20 years of science, Nature Medicine 9 (7) (2003).

8. E. Hernandez-Vargas, R. Middleton, Modeling the three stages in HIV infection, Journal of Theoretical Biology 320 (2013).

9. E. Hernandez-Vargas, Modeling kick-kill strategies toward HIV cure, Frontiers in Immunology 8 (2017).

10. HIV-Causal Collaboration, The effect of combined antiretroviral therapy on the overall mortality of HIV-infected individuals, AIDS, NIH Public Access 24 (1) (2010).

11. E. Sanchez, A. Alanis, A. Loukianov, Discrete-Time High Order Neural Control: Trained with Kalman Filtering, vol. 112, Springer Science & Business Media, 2008.

12. E. Kosmatopoulos, M. Polycarpou, M. Christodoulou, High-order neural network structures for identification of dynamical systems, IEEE Transactions on Neural Networks 6 (2006).

13. A. Alanis, M. Lopez-Franco, N. Arana-Daniel, Discrete-time neural control for electrically driven nonholonomic mobile robots, International Journal of Adaptive Control and Signal Processing 26 (2012).

14. S. Singhal, L. Wu, Training multilayer perceptrons with the extended Kalman algorithm, Advances in Neural Information Processing Systems (1989).

15. S. Haykin, Kalman Filtering and Neural Networks, Wiley Online Library, 2001.

16. S. Saha, G. Raghava, Prediction of continuous B-cell epitopes in an antigen using recurrent neural network, Proteins: Structure, Function, and Bioinformatics 65 (2006).

17. G. Wu, S. Yan, Improvement of prediction of mutation positions in H5N1 hemagglutinins of influenza A virus using neural network with distinguishing of arginine, leucine and serine, Protein and Peptide Letters 14 (2007).

18. X. Qiang, Z. Kou, Prediction of interspecies transmission for avian influenza A virus based on a back-propagation neural network, Mathematical and Computer Modelling 52 (2010).

19. L. Trtica-Majnaric, M. Zekic-Susac, Prediction of influenza vaccination outcome by neural networks and logistic regression, Journal of Biomedical Informatics 43 (5) (2010).

20. G. Hernandez-Mejia, A.Y. Alanis, N. Arana-Daniel, E. Hernandez-Vargas, Recurrent high order neural networks identification for infectious diseases, in: Proceedings in the IEEE World Congress on Computational Intelligence, 2018.

21. G. Rovithakis, M. Christodoulou, Adaptive Control with Recurrent High-Order Neural Networks: Theory and Industrial Applications, Springer Science & Business Media, 2012.

22. Y. Song, J. Grizzle, The extended Kalman filter as a local asymptotic observer for nonlinear discrete-time systems, in: IEEE American Control Conference, 1992.

23. A. Boianelli, V.K. Nguyen, Modeling influenza virus infection: a roadmap for influenza research, Viruses, Multidisciplinary Digital Publishing Institute 7 (2015).

24. B. Agoram, S. Martin, P. Van der Graaf, The role of mechanism-based pharmacokinetic–pharmacodynamic (PK–PD) modelling in translational research of biologics, Drug Discovery Today 12 (2007).

25. P. Van der Graaf, N. Benson, Systems pharmacology: bridging systems biology and pharmacokinetics-pharmacodynamics (PK-PD) in drug discovery and development, Pharmaceutical Research 28 (2011).

26. E.A. Hernandez-Vargas, Modeling and Control of Infectious Diseases: with MATLAB and R, 1st ed., Elsevier Academic Press, 2019.

27. A. Boianelli, N. Sharma-Chawla, D. Bruder, Oseltamivir PK/PD modeling and simulation to evaluate treatment strategies against influenza-pneumococcus coinfection, Frontiers in Cellular and Infection Microbiology 6 (2016).

28. G. Hernandez-Mejia, A. Alanis, E.A. Hernandez-Vargas, Inverse optimal impulsive control based treatment of influenza infection, The 20th World Congress of the International Federation of Automatic Control (IFAC 2017) 50 (2017).

29. A. Alanis, M. Hernandez-Gonzalez, E.A. Hernandez-Vargas, Observers for biological systems, Applied Soft Computing 24 (2014).

CHAPTER 9

Attack Detection and Estimation for Cyber-Physical Systems by Using Learning Methodology

HAIFENG NIU, PHD • C. BHOWMICK, MS • S. JAGANNATHAN, PHD

9.1 INTRODUCTION

Cyber-physical systems, or in particular, networked control systems (NCSs), are the feedback systems where the system components are spatially distributed and connected through communication networks. Examples of such NCSs include cyber-enabled manufacturing, smart grids, and water and sewage networks. The data flow between various parts of the NCS is vulnerable to a variety of potential attacks [26]. Examples of such attacks include jamming attacks [9] and replay attacks [31]. One common attribute of all these attacks, as discussed in [26,23], is that they all tend to deviate the traffic flow in the communication links from the normal value. Inspired by this fact, we propose a traffic flow–based network attack detection and estimation scheme.

Extensive study about flow control has been carried out in the literature [3,8,5]. In [8], a receiver-based flow control scheme is proposed to optimize the amount of data delivered to the destinations via backpressure routing. In [5] a novel max-min flow control scheme using classical sliding mode control is proposed. The work of [3] focuses on joint flow control and scheduling algorithms for multihop wireless networks. However, to the best of our knowledge, minimal effort has been spent on studying flow control from the perspective of network security in the presence of attacks that inject or drop traffic data. Moreover, control of a linear system becomes more challenging in the presence of random disturbances and attacks, especially when the system dynamics are unknown.

On the other hand, the physical system whose feedback loop communication relies on the network links becomes uncertain due to the presence of network delays and packet losses induced by the cyber-attacks [29,15] and can even lead to instability of the overall system. Therefore, it is important to quantify the maximum delays and packet losses that the system can tolerate. In [2], Markov modeling and security measure analysis for networks under denial of service attacks is presented. On the other hand, in [32], the design of robust and resilient controllers is introduced using game theory for CPS with application to power systems. This work was extended in [15] by modeling the CPS and designing a Q-learning scheme. In addition, by adopting an event-triggered control mechanism from [21], which is reported to reduce the data transmission and computation significantly, a stochastic adaptive dynamic programming (ADP)-based approach was employed in [14] to estimate the value function and solve the optimal regulation problem. The performance of the event-triggered control scheme can be improved by utilizing the interevent time to update the unknown parameters, also known as hybrid learning approach [13].

However, the physical system is also subject to attacks; for example, the adversary can manipulate the sensor readings that are being sent to the controller or can inject false data at the actuator [22]. Therefore, it is critical to design a detection scheme for the attacks on the physical system. There are some specific attack models, like replay attacks and deception attacks, which require a sophisticated design of detection schemes. For example, [24] discusses the use of authentication or probing signals in addition to the control input to detect sensor attacks. The work in [11] proposes a novel method to detect replay attacks in a discrete-time linear time-invariant (LTI) system that is controlled using a linear quadratic Gaussian (LQG) optimal controller and equipped with a χ^2 failure detector. However, this approach of detecting physical system attacks [12,10] can only handle zero mean noise, and random disturbances are not considered. It is also important to note that though LQG is an optimal control technique in the presence of noise, the optimality is compromised due to the additional authentication signal. Nevertheless, it is assumed that the system dynamics and the covariance of the noise are known. Here we are interested in designing an attack detection scheme for uncertain systems provided the attack input is unknown but bounded.

Artificial Neural Networks for Engineering Applications. https://doi.org/10.1016/B978-0-12-818247-5.00018-6

In this chapter, the discussion begins with the background of the modeling of NCS and a few typical attack models. Modeling of the communication network from the perspective of flow control is introduced with linear and nonlinear dynamics. Suitable flow controllers are designed assuming the dynamics of the system are uncertain. A finite-time optimal Q-learning algorithm is used in the case of a linear system whereas neural network (NN)-based design approach has been employed in the case of nonlinear dynamics. Residual method–based cyber-attack detection methodologies are discussed and suitable observer designs are presented. The estimation error by using the observer essentially serves as the detection residual and the detectability conditions are also given under which certain types of attacks can be detected by the residual methods. In the residual-based attack detection schemes, one does not have to wait for the system performance to deteriorate; instead, the adversary can be located as soon as the detection signal crosses the threshold. After detection, it is important to estimate the injected attack signal, which could be used to estimate the actual response of the system in the absence of attacks. An attack estimation observer is also presented, and all the proposed algorithms are verified with simulation results.

Linear and nonlinear dynamics are considered for physical systems too. It is mentioned that the physical system dynamics depend on the cyber-states [15] and large delay and packet loss rate from the attacks can cause significant deviation, resulting in instability of the physical system [29]. It is demonstrated that event triggering control saves communication cost and, thereby, reduces the chance of being attacked. For the linear physical system, simultaneous optimization of the control signal and triggering instants is performed using a game-theoretic approach [19]. The ADP-based event-triggered optimal controller is used for the nonlinear physical system, which requires three NNs – identifier, actor, and critic. A physical system attack detection scheme is developed using the identifier residual signal and the performance is demonstrated through simulation.

9.2 BACKGROUND ON SYSTEM MODELING AND ATTACKS

In NCSs, the physical system components are connected through communication networks. In this section, modeling of a typical NCS is discussed by considering the cyber- and the physical side separately. Some typical attack models are also discussed here.

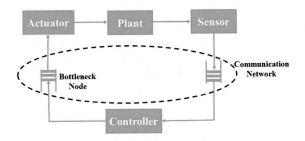

FIG. 9.1 Diagram of a typical NCS.

9.2.1 Modeling

Fig. 9.1 shows the diagram of a typical NCS, in which both controller commands and sensor data are transmitted through a wired or wireless communication link. Here the plant and the communication network can be represented either by a linear or a nonlinear system. The cyber-side, i.e., the dynamics of the network, can be represented as the flow at the bottleneck node.

First the linear representation of the traffic flow model is discussed. Let the input rate at sampling time kT be $\bar{\mu}_k$ packets per second, where $k \in \mathbb{N}$ and T is the sampling period, and $u_k \in \mathbb{R}$ be the adjustment from the previous input rate. Then $\bar{\mu}_k = \bar{\mu}_{k-1} + u_k$. The transmission or service rate $\bar{v}_k$, which slightly fluctuates around the standard transmission rate v_0, is modeled by a stable autoregressive moving average (ARMA) process [1] as $\bar{v}_k = v_0 + \delta_k$, where $\delta_k = \sum_{i=1}^{m} l_i \delta_{k-i} + d_{k-1}$; d_k represents a bounded disturbance with d_M being its bound, m is the number of past values used in the ARMA, and the weights l_i are unknown constants. Let the traffic flow in the bottleneck node at time kT be $\bar{\rho}_k$. Then $\bar{\rho}_{k+1} = \bar{\rho}_k + T\bar{\mu}_k - T\bar{v}_k + w_k$, where w_k is the number of the packets introduced by the attack flow, with $w_k > 0$ and $w_k < 0$ implying injected and dropped data, respectively. Let the shifted flow be $\rho_k = \bar{\rho}_k - \rho_0$ and the shifted input rate be $\mu_k = \bar{\mu}_k - v_0$. Then the augmented flow dynamics become

$$x_{k+1} = Ax_k + Bu_k + Dd_k + Ww_k, \quad (9.1)$$

where $x_k = \begin{bmatrix} \rho_k & \mu_k & \delta_k & \cdots & \delta_{k-m+1} \end{bmatrix}^T \in \mathbb{R}^n$ is the state vector and $A \in \mathbb{R}^{n \times n}$, $B \in \mathbb{R}^n$, $D \in \mathbb{R}^n$, and $W \in \mathbb{R}^n$ with $n \triangleq m + 2$ are the state matrices given by

$$A = \begin{bmatrix} 1 & T & -T & \cdots & 0 & 0 \\ 0 & 1 & 0 & \cdots & 0 & 0 \\ 0 & 0 & l_1 & \cdots & l_{m-1} & l_m \\ 0 & 0 & 1 & \cdots & 0 & 0 \\ \vdots & \vdots & \vdots & \ddots & \vdots & \vdots \\ 0 & 0 & 0 & \cdots & 1 & 0 \end{bmatrix}, B = \begin{bmatrix} 0 \\ 1 \\ 0 \\ 0 \\ \vdots \\ 0 \end{bmatrix},$$

$$D = \begin{bmatrix} 0 \\ 0 \\ 1 \\ 0 \\ \vdots \\ 0 \end{bmatrix}, \quad W = \begin{bmatrix} 1 \\ 0 \\ 0 \\ 0 \\ \vdots \\ 0 \end{bmatrix}.$$

The attack input $w_k \in \mathbb{R}$ is considered unknown but deterministic [6]. Moreover, the system output is defined as

$$y_k = \begin{bmatrix} \tau_k \\ \chi_k \end{bmatrix} = \begin{bmatrix} f_\tau(x_k) \\ f_\chi(x_k) \end{bmatrix} + \begin{bmatrix} v_{\tau,k} \\ v_{\chi,k} \end{bmatrix}, \qquad (9.2)$$

where τ_k is the link end-to-end delay and χ_k is the packet loss rate. The functions $f_\chi : \mathbb{R}^n \to \mathbb{R}$ and $f_\tau : \mathbb{R}^n \to \mathbb{R}$ represent transmission rates whereas $v_{\tau,k}$ and $v_{\chi,k}$ govern the delays and packet loss rate.

Remark 9.1. The network delays and packet losses are primarily determined by the network protocol. The state x_k including input rate, buffer length, and service rate are assumed measurable. However, the delays and packet losses are stochastic variables because they are affected by other factors such as node processing speed, number of hops in the link, and also measurement noise, which are also treated as stochastic variables.

Remark 9.2. Note that matrix A is considered uncertain because $l_1, \cdots, l_m$ are considered unknown, whereas matrices B, D, W are assumed known. A learning-based scheme is required to estimate the uncertain state matrix and control the overall system.

In the case of a nonlinear representation of the communication network, the buffer length at the bottleneck node can be described by the following discrete-time nonlinear system [4]:

$$\begin{aligned} x_{k+1} &= f(x_k) + Tu_k + d_k + w_k, \\ y_k &= \begin{bmatrix} y_{1,k} \\ y_{2,k} \end{bmatrix} = \begin{bmatrix} h_1(x_k) \\ h_2(x_k) \end{bmatrix}, \end{aligned} \qquad (9.3)$$

where $x_k, u_k, d_k, w_k \in \mathbb{R}^n$ are the buffer length, flow input, disturbance input, and attack input at the bottleneck node at time instant kT, respectively; T is the sampling interval, and $f(\cdot)$ denotes the actual traffic accumulation at the bottleneck node. It is a function of buffer length and service capacity, and it is considered unknown. The term y_k represents the system output, where $y_{1,k}$ and $y_{2,k}$ stand for the delay and packet loss, respectively. The relationship between the delay (or packet loss) and the current buffer length is described by the stochastic function h_1 (or h_2). This nonlinear

function is generated using the above model and the information from the network protocol.

The physical system part of the NCS can be linear or nonlinear, depending on the application. A generic system with linear or nonlinear dynamics is considered for the discussion. As the main focus of this study is the security of NCSs, it is important to look into some popular attack models.

9.2.2 Attack Models

This discussion mostly deals with attacks that either inject false data or drop/block authentic data. Three types of such attacks are considered here. The jamming attack aims at creating traffic congestion by placing jammers that consistently inject data into the link. Assuming the attacking strength (number of jammers) increases linearly, this type of attack can be modeled by [25] $w_k = 1 - e^{-\beta k}$, where k, w_k, and β are the time index, percentage of injected data, and attack strength (number of jammers), respectively. The black hole attack compromises one or more nodes in the routing path and hence, part of the data gets discarded. Assuming the attack strength (number of black holes) increases linearly, the black hole attack can be modeled [18] as $w_k = 1 - \beta k$, where k, w_k, and β are the time index, percentage of dropped data, and attack strength (number of black holes), respectively. In the case of a minimum rate DoS streams attack, false data are periodically injected into the network in order to minimize exposure to detection mechanisms. A minimum rate DoS stream attack is given by

$$w_k = \begin{cases} n_1, & \text{for } t \in [k\tilde{T}, k\tilde{T} + p_1] \\ n_2, & \text{for } t \in [k\tilde{T} + p_1, k\tilde{T} + p_2], \\ 0, & \text{for } t \in [k\tilde{T} + p_2, (k+1)\tilde{T}] \end{cases}$$

where n_1, n_2, p_1, p_2, and $\tilde{T}$ are the first attack strength, second attack strength, first attack duration, second attack end time, and total attack period, respectively.

The attacks on the physical system may directly affect the sensors or the actuators. These components can be compromised by the adversary to manipulate the information of the sensors or to alter the control effort at the actuators.

9.3 SECURE LINEAR NETWORKED CONTROL SYSTEMS

As discussed in the previous section, an NCS has two major components. In this section, both the cyber- and physical system dynamics are considered linear and their attack detection schemes are discussed separately.

9.3.1 Network Attack Detection and Estimation

As the linear flow model dynamics are considered unknown, it is of importance to design an adaptive observer that is also utilized for attack detection despite the fact that the state vector is considered measurable. The benefit of the observer is twofold. On the one hand, the unknown flow dynamics A is estimated and on the other hand, the estimation error or residual signal is generated for detection. The observer is described as

$$\hat{x}_{k+1} = A_m \hat{x}_k + (\hat{A}_k - A_m) x_k + B u_k, \qquad (9.4)$$

where $\hat{x}_k \in \mathbb{R}^{n \times 1}$ and $\hat{A}_k \in \mathbb{R}^{n \times n}$ represent the estimated state vector and estimated state matrix, respectively, whereas $A_m \in \mathbb{R}^{n \times n}$ is a design matrix which is selected to be symmetric and Hurwitz. Define the state matrix estimation error as $\tilde{A}_k \triangleq A - \hat{A}_k$ and the state estimation error as $\tilde{x}_k \triangleq x_k - \hat{x}_k$. Then combining (9.4) and (9.1) with $w_k = 0$ yields the state estimation error dynamics as

$$\tilde{x}_{k+1} = A_m \tilde{x}_k + \tilde{A}_k x_k + D d_k. \qquad (9.5)$$

The following assumption is needed before we proceed.

Assumption. The attack is launched after the convergence of the parameter estimation. □

Next we show the performance of the observer in terms of estimating the system parameters, which is summarized in the next theorem.

Theorem 9.1. *(State matrix estimation) Consider the network traffic represented as flow at the bottleneck node described by (9.1). Let the adaptive observer be described by (9.4) with the following update law for the state matrix estimation:*

$$\hat{A}_{k+1} = \hat{A}_k + \beta_1 \frac{\hat{x}_k \tilde{x}_{k+1}^T}{\|\hat{x}_k\|^2 + 1}, \qquad (9.6)$$

where $\beta_1 \in \mathbb{R}^+$ is a design parameter. Then, in the presence of bounded disturbances, $|d_k| \leq d_M$, and the absence of network attacks, $w_k = 0$, both the system matrix estimation error $\tilde{A}_k$ and the state estimation error $\tilde{x}_k$ are uniformly ultimately bounded (UUB).

Remark 9.3. In the absence of disturbances, the state matrix estimation error and the state estimation error will eventually converge to zero.

The network flow represented as an uncertain linear system is controlled by using a Q-learning-based optimal controller. The objective of the controller design is to determine a feedback control policy to minimize the following time-varying finite-horizon value function:

$$J_k = x_N^T S_N x_N + \sum_{i=k}^{N-1} r(x_i, u_i, i)$$

$$\triangleq x_N^T S_N x_N + \sum_{i=k}^{N-1} x_i^T P x_i + u_i^T R u_i, \qquad (9.7)$$

where $P \in \mathbb{R}^{n \times n} \geq 0$, $R \in \mathbb{R}^+$ are the penalty matrices, $N < \infty$ represents the final time, $S_N \in \mathbb{R}^{n \times n} \geq 0$ is the penalty matrix for x_N, the state vector at the final time instant, and $r(x_i, u_i, i)$ represents the cost-to-go from time instant i onward, which is a function of x_i, the state vector at the time instant i, and u_i, the control input at the time instant i. It is known that the finite-horizon optimal control input u_k^* can be found by solving the Riccati equation [7]. However, the Riccati equation cannot be solved in this case since the system dynamics are unknown. We use a finite-time Q-learning approach here [30].

Define the optimal action–dependent value function Q as

$$Q(x_k, u_k, N - k) = r(x_k, u_k, k) + J_{k+1}$$

$$\triangleq \begin{bmatrix} x_k \\ u_k \end{bmatrix}^T G_k \begin{bmatrix} x_k \\ u_k \end{bmatrix}, \qquad (9.8)$$

where J_{k+1} is the value function at time instant $k + 1$ and the matrix G_k is defined as

$$G_k = \begin{bmatrix} P + A^T S_{k+1} A & A^T S_{k+1} B \\ B^T S_{k+1} A & R + B^T S_{k+1} B \end{bmatrix}$$

$$\triangleq \begin{bmatrix} G_k^{xx} & G_k^{xu} \\ G_k^{ux} & G_k^{uu} \end{bmatrix}, \qquad (9.9)$$

with S_{k+1} being the solution to Riccati equation at time instant $k + 1$. The elements of G_k are given as $G_k^{xx} = P + A^T S_{k+1} A$, $G_k^{xu} = A^T S_{k+1} B$, $G_k^{ux} = B^T S_{k+1} A$, and $G_k^{uu} = R + B^T S_{k+1} B$. The optimal control input is

$$u_k^* = (G_k^{uu})^{-1} G_k^{ux} x_k. \qquad (9.10)$$

The control input can be computed from G_k, when the system dynamics are known.

Assumption. [30] The slowly time-varying Q-function can be expressed as linear in the unknown parameters (LIP). □

With the above Assumption, we express $Q(x_k, u_k, N - k)$ as

$$Q(x_k, u_k, N - k) = z_k^T G_k z_k \triangleq g_k^T \bar{z}_k, \qquad (9.11)$$

where $z_k \triangleq \begin{bmatrix} x_k^T & u_k^T \end{bmatrix}^T$ is the augmented state vector, $\bar{z}_k$ is the Kronecker product quadratic polynomial basis vector of z_k, and g_k is a vector generated by stacking the columns of G_k into a one-column vector with the summed off-diagonal elements. Now

$$g_k = \phi(N - k)\theta_k,$$

where $\theta_k \in \mathbb{R}^{L \times 1}$ with $L = \frac{n(n+1)}{2}$ is the target parameter vector and $\phi(N - k)$ is the basis function matrix defined as $[\phi(N - k)]_{i,j} = exp\left(-tanh(N - k)^{L+1-j}\right)$. The estimated value of g_k is given by $\hat{g}_k = \phi(N - k)\hat{\theta}_k$, where $\hat{\theta}_k$ is the estimated value of the target parameter vector θ_k. The estimated Q-function follows the same structure as Q and is given by

$$\hat{Q}(x_k, u_k, N - k) = \hat{g}_k^T \bar{z}_k = \hat{\theta}_k^T \phi(N - k)^T \bar{z}_k \triangleq \hat{\theta}_k^T \check{z}_k, \tag{9.12}$$

where $\check{z}_k = \phi(N - k)^T \bar{z}_k$ is the regression function satisfying $\check{z}_k = 0$ for $\bar{z}_k = 0$. Accordingly, the actual control input becomes

$$u_k = \left(\hat{G}_k^{uu}\right)^{-1} \hat{G}_k^{ux} x_k, \tag{9.13}$$

where $\hat{G}_{uu}$ and $\hat{G}_{ux}$ are the elements of $\hat{G}$, which is given as

$$\hat{G} \triangleq \begin{bmatrix} \hat{G}_{xx} & \hat{G}_{xu} \\ \hat{G}_{ux} & \hat{G}_{uu} \end{bmatrix}$$
$$= \begin{bmatrix} P + \hat{A}_k^T \hat{S}_{k+1} \hat{A}_k & \hat{A}_k^T \hat{S}_{k+1} B \\ B^T \hat{S}_{k+1} \hat{A}_k & R + B^T \hat{S}_{k+1} B \end{bmatrix},$$

where $\hat{S}_{k+1}$ is the estimate of the solution to the Riccati equation at the time instant $k + 1$. In Q-learning, $\hat{g}_k$ is updated by the update law of the parameter vector $\hat{\theta}_k$, so $\hat{G}_k$ is formed by reshaping $\hat{g}_k$ from a vector to a matrix of proper dimension.

Note that if $\hat{G}_k^{uu}$ is singular, then it is replaced by R. By using the adaptive observer (9.4), the Bellman error is given by

$$e_{b,k+1}$$
$$= \begin{bmatrix} \hat{x}_k \\ u_k \end{bmatrix}^T \begin{bmatrix} P + \hat{A}_k^T \hat{S}_{k+1} \hat{A}_k & \hat{A}_k^T \hat{S}_{k+1} B \\ B^T \hat{S}_{k+1} \hat{A}_k & R + B^T \hat{S}_{k+1} B \end{bmatrix} \begin{bmatrix} \hat{x}_k \\ u_k \end{bmatrix}$$
$$- \hat{x}_k^T \hat{S}_k \hat{x}_k + 2 \left(\hat{A}_k \hat{x}_k + B u_k\right)^T \hat{S}_{k+1} A_m \tilde{x}_k \tag{9.14}$$
$$- \tilde{x}_k^T A_m^T \hat{S}_{k+1} A_m \tilde{x}_k + \left(\hat{\theta}_k^T \varphi(0) - g_N\right) I_v$$
$$= -\hat{\theta}_k^T \Delta \check{z}_k + r\left(\hat{x}_k, u_k, k\right) - \tilde{\theta}_k^T \varphi(0) I_v,$$

where $\Delta \check{z}_k \triangleq \check{z}_k - \check{z}_{k-1}$ and $I_v \triangleq \begin{bmatrix} 1 & \cdots & 1 \end{bmatrix}^T \in \mathbb{R}^{L \times 1}$. Let $\sigma_k \triangleq \Delta \check{z}_k - \phi(0) I_v$, $\bar{\varphi}(x_k) \triangleq kron([x_k, u_k])$, $\bar{\varphi}(\hat{x}_k) \triangleq kron([\hat{x}_k, u_k])$ with $kron(\cdot)$ being the quadratic polynomial of the Kronecker product. Then (9.14) can be rewritten as

$$e_{b,k+1} = \tilde{\theta}_k^T \sigma_k + \hat{\theta}^T \left(\bar{\varphi}(x_k) - \bar{\varphi}(\hat{x}_k)\right) + r(x_k, u_k, k) - r(\hat{x}_k, u_k, k). \tag{9.15}$$

The update law for the value function estimator is given by

$$\hat{\theta}_{k+1} = \hat{\theta}_k + \beta_2 \frac{\sigma_k e_{b,k+1}}{\|\sigma_k\|^2 + 1}, \tag{9.16}$$

where $\beta_2 > 0$ is the tuning rate. Then the parameter error dynamics for $\tilde{\theta}_k$ becomes

$$\tilde{\theta}_{k+1} = \tilde{\theta}_k - \beta_2 \frac{\sigma_k e_{b,k+1}}{\|\sigma_k\|^2 + 1}. \tag{9.17}$$

Using stability analysis by the Lyapunov method, it can be shown that in the presence of bounded disturbance and in the absence of network attacks, the signals $x, \tilde{x}, \tilde{A}, \tilde{\theta}$ are all UUB, when $\hat{\theta}$ is updated using (9.16) with $0 < \beta < 1/5$.

Now the attack detection scheme based on observer (9.4) is discussed. The detectability condition is derived under which certain attacks can be detected. An observer is proposed to estimate the attack flow.

Theorem 9.2. (*Network attack detectability condition* [14]) *Consider the network traffic represented as a flow at the bottleneck node described by* (9.1). *Let the adaptive observer be described by* (9.4) *with the update law given by* (9.6). *Let the control input be given by* (9.13) *with the estimated value function tuned by* (9.16). *Assume the attack is launched after the convergence of the system states* x_k, *system matrix estimation error* $\tilde{A}_k$, *and state estimation error* $\tilde{x}_k$ *at time* $t = k_0 T + \bar{N}$ *with* $\bar{N} \geq N \in \mathbb{N}$ *and* k_0 *being the initial time. Here* T *represents the sampling time. Then the attack is detectable for all time* $t \geq k_0 T + \bar{N}$ *if the injected (dropped) traffic flow* w_k *into (from) the link satisfies*

$$\left\| \sum_{i=k_0}^{k-1} A_m^{k-k_0-i-1} W w_i \right\|$$
$$> \Upsilon_k + \left\| \sum_{i=k_0}^{k-1} A_m^{k-k_0-i-1} (D d_M + \tilde{A}_i x_i) \right\|, \tag{9.18}$$

where Υ_k *is the bound for* $\|\tilde{x}_k\|$ *in the healthy condition.*

Upon detecting the attack, it is of interest to know how much flow has been injected or dropped by the attack. Modifying (9.4), the attack detection observer is defined as

$$\hat{x}_{k+1} = A_m \hat{x}_k + (\hat{A}_k - A_m)x_k + Bu_k + W\hat{w}_k, \quad (9.19)$$

where $\hat{w}_k$ is the estimated attack flow. The update law for the estimated attack flow is defined as

$$\hat{w}_k = \hat{w}_{k-1} + \alpha_3 \tilde{x}_{k-1}^T \bar{K} \hat{A}^T W - \alpha_4 \left| 1 - \alpha_3 W^T W \right| \hat{w}_{k-1}, \quad (9.20)$$

where $\alpha_3, \alpha_4 \in \mathbb{R}^+$ and $\bar{K} \in \mathbb{R}^{n \times n}$ are design parameters. The following theorem guarantees the boundedness of the state estimation error $\tilde{x}_k$ and the attack flow estimation error $\tilde{w}_k = w_k - \hat{w}_k$.

Theorem 9.3. *(Closed-loop stability with attacks) Consider the closed-loop system (9.1) and observer (9.19) with state matrix identifier update (9.6) and estimated attack flow update (9.20). Let the disturbance bound be d_M and the attack input satisfy $\|w_k\| \le w_M$. Suppose the attacks can be detected when the network detection residual exceeds a predefined threshold given by Υ_k. Then the network attack residual $\tilde{x}_k$ and the estimation error of the attack flow $\tilde{w}_k$ are UUB.*

Remark 9.4. The attack detection and estimation scheme for the network part are discussed in standard discrete-time domain. But in practice, the network and the physical system are typically event-driven due to the presence of the communication network, and one can use event-triggered control. In the next subsection, the physical system control design using an event-triggered mechanism is discussed and one can use the same event triggering condition for the network as well.

9.3.2 Attack Detection and Estimation in Linear Physical System

It was derived in [29] that a linear time-invariant system after the incorporation of random delays and packet losses becomes an uncertain stochastic time-varying system. This stochastic system is considered for attack detection. Consider the stochastic linear continuous-time system with network-induced delays and packet losses described by

$$\dot{x}_p(t) = A_p x_p(t) + \chi(t) B_p u_p(t - \tau(t)) \\ + \chi'(t) w_p(t - \tau'(t)), \quad (9.21)$$

where $x_p(t) \in \mathbb{R}^n$, $u_p(t) \in \mathbb{R}^m$, and $w_p(t) \in \mathbb{R}^n$ are the system states, controller input, and attack flow input

vector, respectively, $A_p \in \mathbb{R}^{n \times n}$ and $B_p \in \mathbb{R}^{n \times m}$ are the system matrices which are considered unknown, $0 \le \tau \le \tau_M$ and $0 \le \tau' \le \tau'_M$ stand for the network-induced delays which are considered bounded, and $\chi(t)$ and $\chi'(t)$ are the packet loss indicators which are equal to the identity matrix I_n when the packet is received and the null matrix when the packet is lost.

Remark 9.5. The term $w_p(t)$ is used to characterize the change in system state vector caused by attacks on the sensors or actuators [17]. From the NCS diagram in Fig. 9.1, it can be seen that if $w_p(t)$ is the sensor attack input, then we have $\tau'(t) = \tau(t)$ and $\chi'(t) = \chi(t)$. On the other hand, if $w_p(t)$ is the actuator attack input, we will have $\tau'(t) = 0$ and $\chi' = I_n$ since the actuator-plant link does not rely on the network.

Remark 9.6. It is important to note that although it appears from (9.21) that the attack affects the system states, the representation (9.21) is not limited to the case where the attack targets the state vector [17]. For instance, in the case of actuator attacks the controller input is manipulated from $u_{p,k}$ to $u_{p,k} + \Delta u_{p,k}$, where $\Delta u_{p,k}$ is the change in input signal, and the attack input term in (9.21) becomes $B_p \Delta u_{p,k}$. Likewise, for sensor attack, $x_{p,k}$ is manipulated to $x_{p,k} + \Delta x_{p,k}$, where $\Delta x_{p,k}$ is the change in sensor data and the system dynamics become $x_{p,k+1} = A_p x_{p,k} + B_p K_{p,k} (x_{p,k} + \Delta x_{p,k})$, so the attack input is $w_{p,k} = B_p K_{p,k} \Delta x_{p,k}$.

While designing the optimal control, it is assumed initially that there is no attack on the physical system. The subscript p is now dropped for simplification. The continuous-time stochastic system dynamics in the absence of attacks, but in presence of delays and packet losses can be found from (9.21) as

$$\dot{x}(t) = A_p x(t) + \chi(t) B_p u(t - \tau(t)). \quad (9.22)$$

The following assumption is needed before we proceed.

Assumption. [21] (A_p, B_p) is controllable and the states are measurable. The initial value of the state vector is deterministic. The time-varying delays are random but bounded and the packet losses are i.i.d. and follow a Bernoulli distribution. The maximum delay bound is d sampling instants, where d is a positive integer. $\quad \square$

In this event-triggered control, define the transmission instants as $\{k_i\}_{i=0}^{\infty}$ of $k \in \mathbb{N}$ with initial time $k_0 = 0$. The state vector received at the controller x_k^s is given by

$$x_k^s = x_{k_i}, \quad k_i \le k < k_{i+1}, \quad i = 1, 2, \cdots. \quad (9.23)$$

The event-based control input u_k^s at the controller is given by

$$u_k^s = \mu(x_k^s) = \mu(x_{k_i}), \quad k_i \le k < k_{i+1}, \quad i = 1, 2, \cdots, \tag{9.24}$$

where $\mu : \mathbb{R}^n \to \mathbb{R}^m$ is the mapping from the system states to the control input signal. The event trigger error is defined as

$$e_k^u = u_k^s - u_k = \mu(x_k^s) - \mu(x_k). \tag{9.25}$$

The event-based discrete-time equivalent of (9.22) after incorporating the delays, packet losses, and event-based current and delayed control inputs is given by

$$x_{k+1} = A_p^s x_k + \sum_{l=0}^{d} \chi_{k-l} B_{p,l}^k u_{k-l} + \sum_{l=0}^{d} \chi_{k-l} B_{p,l}^k e_{k-l}^u, \tag{9.26}$$

where $x_k = x(kT)$ are the states of the discrete-time system and $A_p^s = e^{AT}$ and $B_{p,l}^k = \int_{t_l^k}^{t_{l-1}^k} e^{A_p(T-w)} dw B_p$ are the system matrices with $t_{-1}^k = T$ and $t_d^k = 0$. Defining the augmented state as

$$z_k = \begin{bmatrix} x_k^T & u_{k-1}^{uT} & \cdots & u_{k-d}^{uT} & e_{k-1}^{uT} & \cdots & e_{k-d}^{uT} \end{bmatrix}^T \in \mathbb{R}^{n+2dm},$$

the discretized system (9.26) can be expressed as

$$z_{k+1} = A_{p,k}^{\tau,\chi} z_k + B_{p,k}^{\tau,\chi} u_k + E_{p,k}^{\tau,\chi} e_k^u, \tag{9.27}$$

where $A_{p,k}^{\tau,\chi} \in \mathbb{R}^{(n+2dm)\times(n+2dm)}$, $B_{p,k}^{\tau,\chi} \in \mathbb{R}^{(n+2dm)\times m}$, $E_{p,k}^{\tau,\chi} \in \mathbb{R}^{(n+2dm)\times m}$ are the stochastic time-varying matrices, defined in [21], and z_k^s will be used to denote the event-based augmented state at time instant k using the event-based states x_k^s, the event-based control input u_k^s, and the event trigger error e_k^u. Define the following cost function for the cooptimization as

$$J(z_k) = \mathbb{E}_{\tau,\chi} \left\{ \sum_{j=k}^{\infty} r(z_j, u_j, e_j^u) | z_k \right\}, \tag{9.28}$$

where $r(z_k, u_k, e_k^u) = z_k^T H_z z_k + u_k^T R_z u_k - \gamma_z^2 e_k^{uT} e_k^u$ is the cost-to-go from time instant k onward, with $H_z = diag\{H, \frac{R}{d+1}, \cdots, \frac{R}{d+1}\}$, $R_z = \frac{R}{d+1}$, $\gamma_z^2 = \frac{\gamma^2}{d+1}$. Here $H \in \mathbb{R}^{n\times n}$ and $R \in \mathbb{R}^{m\times m}$ are constant symmetric positive definite penalty matrices for x_k and u_k, respectively, and

$\gamma > \gamma^*$ is the penalty of control input error, e_k^u, as defined in [21]. The optimization problem can be represented as a min-max problem where the optimal value $V^*(z_k)$ of the cost function (9.28) is given by

$$V^*(z_k) = \min_{u_k} \max_{e_k^u} \mathbb{E}_{\tau,\chi} \left\{ \sum_{j=k}^{\infty} r(z_j, u_j, e_j^u) + V(z_{k+1}) | z_k \right\},$$

where $V(z_{k+1})$ is the cost-to-go from time $k + 1$ onward. Assuming that the stochastic game has a solution, the stochastic optimal value function $V^*(z_k)$ can be expressed in a quadratic form as

$$V^*(z_k) = \mathbb{E}_{\tau,\chi} \{z_k^T S_k z_k\}, \tag{9.29}$$

where S_k is the positive definite solution to the stochastic game Riccati (SGR) equation [21]. The optimal control input and the optimal control input error are denoted with a superscript $*$ and are given as $u_k^* = K_k^* z_k$, $e_k^{u*} = L_k^* z_k$, where K_k^* and L_k^* are optimal gains as defined in [21]. As the system dynamics are unknown, the optimal gains cannot be computed directly; hence, stochastic Q-learning approach is used. The Q-function can be written as

$$Q(z_k, u_k, e_k^u) = \mathbb{E}_{\tau,\chi} \left\{ r(z_k, u_k, e_k^u) + V^*(z_{k+1}) | z_k \right\}$$

$$= \mathbb{E}_{\tau,\chi} \left\{ \begin{bmatrix} z_k^T & u_k^T & e_k^{uT} \end{bmatrix} G_k \begin{bmatrix} z_k^T & u_k^T & e_k^{uT} \end{bmatrix}^T \right\} \tag{9.30}$$

$$= \mathbb{E}_{\tau,\chi} \left\{ \bar{\eta}_k^T G_k \bar{\eta}_k \right\},$$

where $\bar{\eta}_k = \begin{bmatrix} z_k^T & u_k^T & e_k^{uT} \end{bmatrix}^T \in \mathbb{R}^{n+(d+2)m \triangleq l_{mn}}$ is the augmented state and $G_k \in \mathbb{R}^{l_{mn}\times l_{mn}}$ is defined as follows:

$$G_k = \begin{bmatrix} G_k^{zz} & G_k^{zu} & G_k^{ze} \\ G_k^{uz} & G_k^{uu} & G_k^{ue} \\ G_k^{ez} & G_k^{eu} & G_k^{ee} \end{bmatrix}, \tag{9.31}$$

where the expression for each individual element can be found in [21]. The optimal control input gain K_k^* and optimal control input error gain L_k^* are

$$K_k^* = (G_k^{uu} - G_k^{ue} G_k^{ee-1} G_k^{eu})^{-1} (G_k^{ue} G_k^{ee-1} G_k^{ez} - G_k^{uz}),$$

$$L_k^* = (G_k^{eu} - G_k^{eu} G_k^{uu-1} G_k^{ud})^{-1} (G_k^{ue} G_k^{ee-1} G_k^{ez} - G_k^{uz}). \tag{9.32}$$

Assumption. [21] The Q-function is slowly time-varying and can be expressed as LIP. □

Therefore, the Q-function can be expressed as

$$Q(z_k, u_k, e_k^u) = \mathop{\mathbb{E}}_{\tau, \chi}\{g_k^T \xi_k\}, \quad \forall k, \qquad (9.33)$$

where the parameter vector $g_k \in \mathbb{R}^{l_{mn}^2}$ is the vectorized form of matrix G_k and $\xi_k = \eta_k \otimes \eta_k \in \mathbb{R}^{l_{mn}^2}$ is the quadratic polynomial regression vector. The Q-function estimator (QFE) $\hat{Q}$ is represented as

$$\hat{Q}(z_k^s, u_k^s, e_k^u) = \mathop{\mathbb{E}}_{\tau, \chi}\{\hat{g}_k^T \xi_k^s\}, \quad k_i \leq k < k_{i+1}, \qquad (9.34)$$

where $\hat{Q}$ and $\hat{g}_k$ are the estimates of the Q-function and g_k, respectively; $\xi_k^s = \eta_k^s \otimes \eta_k^s$ is the event-based quadratic polynomial regression vector at time instant k, with $\eta_k^s = \left[z_k^{s^T} \; u_k^{s^T} \; e_k^{u^T} \right]^T$. The estimated control gain matrix $\hat{K}_k$ and estimated control input error gain matrix $\hat{L}_k$ are given by

$$\hat{K}_k = (\hat{G}_k^{uu} - \hat{G}_k^{ue}\hat{G}_k^{ee^{-1}}\hat{G}_k^{eu})^{-1}(\hat{G}_k^{ue} - \hat{G}_k^{ez}\hat{G}_k^{ee^{-1}}\hat{G}_k^{uz}),$$
$$\hat{L}_k = (\hat{G}_k^{eu} - \hat{G}_k^{eu}\hat{G}_k^{uu^{-1}}\hat{G}_k^{ud})^{-1}(\hat{G}_k^{eu} - \hat{G}_k^{uu^{-1}}\hat{G}_k^{uz}\hat{G}_k^{ez}),$$
$$(9.35)$$

where $\hat{G}_k$ is the estimate of G_k and $\hat{G}_k^{(\cdot)}$ denotes the estimate of $G_k^{(\cdot)}$. The event-based estimated control input and estimated control input error are computed as

$$u_k^s = \hat{K}_k z_k^s, \quad k_i \leq k < k_{i+1},$$
$$e_k^u = \hat{L}_k z_k^s, \quad k_i \leq k < k_{i+1}. \qquad (9.36)$$

With estimated $\hat{Q}$, the Bellman error is represented as

$$e_k^{s,B} = \mathop{\mathbb{E}}_{\tau, \chi}\{r^s(z_{k-1}^s, u_{k-1}^s, e_{k-1}^u)\}$$
$$+ \mathop{\mathbb{E}}_{\tau, \chi}\{\hat{g}_k^T \Delta\xi_{k-1}^s | z_{k_i}, z_{k_i-1}\}, \quad k_i \leq k < k_{i+1}, \qquad (9.37)$$

where $r^s(z_{k-1}^s, u_{k-1}^s, e_{k-1}^u) = r(z_{k_i-1}, u_{k_i-1}, e_{k_i-1}^u)$ is the event-based cost-to-go from time instant $k - 1$ onward and $\Delta\xi_{k-1}^s = \xi_{k_i}^s - \xi_{k-1}^s, k_i \leq k < k_{i+1}$. The recursive least square update for the estimated Q-function parameter is given by

$$\hat{g}_{k_i+j} = \mathop{\mathbb{E}}_{\tau, \chi}\left\{ \hat{g}_{k_i-1+j} - \frac{\alpha_g W_{k_i-2+j}\Delta\xi_{k_i-1}^s e_{k_i-1+j}^{s,B^T}}{1 + \Delta\xi_{k_i-1}^{s^T} W_{k_i-2+j}\Delta\xi_{k_i-1}^s} \right\},$$
$$j = 0, 1, ..., k_{i+1} - k_i - 1, \qquad (9.38)$$

where the Bellman error dynamics for $k_i + j, j = 0, 1, 2, \cdots, k_{i+1} - k_i - 1$, are represented as

$$e_{k_i-1+j}^{s,B} = \mathop{\mathbb{E}}_{\tau, \chi}\{r^s(z_{k_i-1}, u_{k_i-1}, e_{k_i-1}^u)\}$$
$$+ \mathop{\mathbb{E}}_{\tau, \chi}\{\hat{g}_{k_i-1+j}^T \Delta\xi_{k_i-1}^s\}, \quad j = 0, 1, 2, ..., k_{i+1} - k_i - 1, \qquad (9.39)$$

and the dynamics of the positive definite gain matrix W are given by

$$W_{k_i+j}$$
$$= \left(W_{k_i-1+j} - \alpha_w \frac{W_{k_i-1+j}\Delta\xi_{k_i-1}^s \Delta\xi_{k_i-1}^{s^T} W_{k_i-1+j}}{1 + \Delta\xi_{k_i-1}^{s^T} W_{k_i-1+j}\Delta\xi_{k_i-1}^s} \right)$$
$$\times \mathbf{1}_{\{||W_{k_i+j}||\geq W_{min}\}},$$
$$j = 0, 1, 2, ..., k_{i+1} - k_i + 1, \qquad (9.40)$$

where $W_0 = \beta_w I$, $\beta_w > 0, \alpha_g > 0, \alpha_w > 0$ are constants and W_{min} is the minimum value of the gain matrix indicated by indicator function $\mathbf{1}_{(\cdot)}$. The update law is defined for all time instants; in particular, update during the time between the trigger instants accelerates the parameter update. It is shown in [21] that when the Q-function parameters are initialized with nonzero values in a compact set and the parameters are updated using (9.38) and (9.40), the closed-loop event-triggered system is asymptotically stable in the mean square provided the regression vector satisfies PE, the inequality

$$e_k^{u^T} e_k^u \leq \mathop{\mathbb{E}}_{\tau, \chi}\{z_k^{s^T} \hat{L}_k^T \hat{L}_k z_k^s\}, \quad k_i \leq k < k_{i+1}, \qquad (9.41)$$

holds, and the parameters α_g, γ_z satisfy certain bounds. Moreover, the control law and the transmission instants achieve optimality.

In the above analysis, the network is assumed to be in a healthy condition, i.e., the delays and packet losses are bounded by a small value. They increase in the presence of attacks on the network and hence, it is of interest to determine the maximum delays and packet losses that the physical system can tolerate.

Let $[k_l, k_l + \epsilon_{p,k}]$ be the interval during which there are no sensor data received at the controller; $\epsilon_{p,k}$ depends on the event trigger error, network-induced delays, and packet losses. This can be explained with the following example. Suppose the event is triggered at $k_l = 0$ and the controller received the event with no delay. The next event is triggered at $k_{l+1} = 3$; however, the packet containing this event is lost. Then the event will be triggered again at $k = 4$, since the control input has not been changed and the trigger error keeps increasing. Suppose that the network-induced delay is $\tau = 2T$, where T denotes the sampling time. Then the time that the controller receives the event will be $k = 6$. Therefore, in this case we have $\epsilon_{p,k} = 6T$. The following theorem gives the maximum time span $\epsilon_{p,k}$ that the physical system can tolerate.

Theorem 9.4. *(Maximum tolerable delay [14]) Consider the system (9.27) without physical attacks with control input and control input error given in (9.36). Suppose the estimated Q-function is expressed as (9.34) with update laws given by (9.38) and (9.40). Assume that the communication network is under attacks such that the time span $\epsilon_{p,k}$ is always greater than ϵ_m, which denotes the maximum tolerable delay for the system. Then the physical system becomes unstable if ϵ_m satisfies*

$$
\lambda_{min}\{\Xi\} \mathop{\mathbb{E}}_{\tau,\chi} \left\{ \left\| \begin{array}{l} (A_{p,k}^{\tau,\chi})^{\epsilon_m} z_{k_l} + \sum_{i=k_l}^{k_l+\epsilon_m-1} (A_{p,k}^{\tau,\chi})^{k_l+\epsilon_m-1} B_{p,k}^{\tau,\chi} u_i \\ + \sum_{i=k_l}^{k_l+\epsilon_m-1} (A_{p,k}^{\tau,\chi})^{k_l+\epsilon_m-1} E_{p,k}^{\tau,\chi} e_i^u \end{array} \right\|^2 \right\} \geq
$$

$$
\lambda_{max}\{\Xi\} \mathop{\mathbb{E}}_{\tau,\chi} \left\{ \|z_{k_l}\|^2 \right\} + \xi_\delta^2 (\alpha_g \xi_{min}^2 (2 - \alpha_g)) \| \mathop{\mathbb{E}}_{\tau,\chi} \{\tilde{g}_k\} \|^2, \tag{9.42}
$$

where $\mathbb{E}\{\tilde{g}_k\} = \mathbb{E}\{g_k\} - \mathbb{E}\{\hat{g}_k\}$ denotes the Q-function parameter estimation error and $\Xi \in \mathbb{R}^{(n+2dm) \times (n+2dm)}$ is a positive definite symmetric matrix and $\xi_\delta > 0$ is a constant, both defined in [21].

Remark 9.7. Theorem 9.4 gives the maximum tolerable delay resulting from network-induced delay and packet losses that the physical system can tolerate before it may become unstable. Appropriate network defense must be launched or the system needs to be shut down once $\epsilon_{p,k}$ exceeds this threshold.

Next, it is assumed that the network is in the healthy condition whereas the physical system suffers from attacks. The detectability condition is derived under which the attacks on the physical system can be detected. Consider the system in the presence of physical attacks as

$$
z_{k+1} = A_{p,k}^{\tau,\chi} z_k + B_{p,k}^{\tau,\chi} u_k + E_{p,k}^{\tau,\chi} e_k^u + W_{p,k} \omega_k, \tag{9.43}
$$

where ω_k denotes the unknown but bounded discrete-time attack input signal and $W_{p,k}$ is the attack input matrix. Suppose that ω_k is considered as a disturbance and no defenses will be launched, if it is smaller than a

given threshold, i.e., $\|\omega_k\| \leq \omega_{pM}$. The following theorem shows the boundedness of the system state vector when $\|\omega_k\| \leq \omega_{pM}$.

Theorem 9.5. *(Attack detection on the physical system [14]) Consider the system (9.43) in the absence of attacks on the network and let the control input and control input error be given by (9.36). Let u_0 be an initial admissible control input. The estimated Q-function is given in (9.34) with update laws (9.38) and (9.40). The events are triggered when the inequality in (9.41) does not hold. Assume the regression vector satisfies the PE condition. Let ω_{pM} be the threshold below which the attack input ω_k would be considered as a disturbance. Then the attack can be detected when the system state vector satisfies $\mathop{\mathbb{E}}_{\tau,\chi} \{\|z_k\|\} > \tilde{\Upsilon}_k$, where $\tilde{\Upsilon}_k$ is considered as a threshold to detect attacks and is an upper bound to the norm of the states derived from its behavior in the healthy system condition.*

9.4 SECURE NONLINEAR NETWORKED CONTROL SYSTEMS

In the previous section, attack detection and estimation schemes are given when the network and the physical system are viewed as linear systems. Now we will extend the residual-based approach for nonlinear networked control systems. A few of the mathematical symbols used in this section may have been used in Section 9.3 to denote other variables for convenience and to follow standardized notions. However, the variables used in this section will be defined and the readers are advised not to make any reference for them to the previous section.

9.4.1 Nonlinear Network Attack Detection and Estimation

To design the controller, let the desired buffer length at time instant k be $x_{d,k}$ and define the error in buffer length as

$$
e_k = x_k - x_{d,k}, \tag{9.44}
$$

where x_k denotes the actual buffer length at the bottleneck node at time instant k as defined in (9.3). The objective is to derive a suitable input flow rate u_k such that the error between the desired and actual buffer length can be minimized. Using definition of error (9.44) in the state dynamics (9.3), we have the tracking error dynamics as

$$
e_{k+1} = f(x_k) + T u_k + d_k + w_k - x_{d,k+1}, \tag{9.45}
$$

where d_k and w_k denote the disturbance input and attack input at time instant k, as defined in (9.3), and T

denotes the sampling interval. Since the nonlinear function $f(\cdot)$ is unknown, an one-layer NN will be utilized to estimate $f(\cdot)$. Let

$$f(x_k) = \theta_k^T \varphi(x_k) + \varepsilon_k, \qquad (9.46)$$

where $\theta \in \mathbb{R}^{L \times 1}$ is a vector of constant target weights bounded by $\|\theta\| \le \theta_M$, which are assumed to be unknown, $\varphi(\cdot) \in \mathbb{R}^{L \times 1}$ is the activation function bounded by $\varphi(\cdot) \le \varphi_M$, L being the number of neurons used in the NN, and ε_k is the NN functional construction error vector. Define the NN output as

$$\hat{f}(x_k) = \hat{\theta}_k^T \varphi(x_k), \qquad (9.47)$$

where $\hat{\theta}_k \in \mathbb{R}^{L \times 1}$ are the weights of the NN. The control input u_k is designed as

$$u_k = \left(-\hat{f}(x_k) + x_{d,k+1} + Ke_k \right) / T, \qquad (9.48)$$

where K is a diagonal feedback gain matrix for the control input [4]. Substituting (9.48) into (9.45) yields the closed-loop buffer length error dynamics

$$e_{k+1} = Ke_k + \tilde{\theta}_k^T \varphi(x_k) + \varepsilon_k + d_k + w_k, \qquad (9.49)$$

where $\tilde{\theta}_k = \theta_k - \hat{\theta}_k$ is the parameter error during the estimation. From (9.49) it can be seen that the buffer length error depends on the parameter error, disturbance input, and attack input.

Assumption. The desired buffer length $x_{d,k}$ is bounded and the NN reconstruction error ϵ_k is bounded by a known constant [4]. The disturbance input and the attack input are bounded as $\|d\| \le d_M$, $\|w\| \le w_M$, where d_M, w_M are known constants. ☐

It is shown in [4] that an NN controller stabilizes the buffer length to a desired level in the absence of an attack. Next, an adaptive NN observer is presented to show that the buffer length estimation error and the parameter estimation error converge to a small subset. It has been reported in the literature [1] that the states of the communication network can be easily measured when the servers at the output queues are rate allocating servers and the transport protocol supports the packet-pair probing technique. In this chapter, the network states described by the input rate and the current buffer length in the link are considered accessible. Next, we will present the novel observer and show that the estimation and the modeling parameter errors converge to a small subset in the absence of attack input.

Let $\hat{x}_k$ be the estimated buffer length from the observer. The dynamics of the observer is given by the following linear time-varying equation:

$$\hat{x}_{k+1} = \hat{\theta}_k^T \varphi(x_k) + Tu_k - L(x_k - \hat{x}_k), \qquad (9.50)$$

where $L \in \mathbb{R}$ is the observer feedback gain matrix. Define the buffer length estimation error as $\tilde{x}_k = x_k - \hat{x}_k$. Then, combining (9.50) and (9.3) yields the buffer length estimation error dynamics as

$$\tilde{x}_{k+1} = L\tilde{x}_k + \tilde{\theta}_k^T \varphi(x_k) + \varepsilon_k + d_k + w_k. \qquad (9.51)$$

The parameter $\hat{\theta}_k$ is updated using the following update law [16]:

$$\hat{\theta}_{k+1} = \hat{\theta}_k + \alpha \varphi(x_k) \tilde{x}_{k+1}^T - \Gamma \left\| I - \alpha \varphi(x_k) \varphi^T(x_k) \right\| \hat{\theta}_k, \qquad (9.52)$$

where $\alpha \in \mathbb{R}^+$, $\Gamma \in \mathbb{R}^+$ are design parameters. It has been shown in [16] that the state estimation error $\tilde{x}$ and the parameter estimation error $\tilde{\theta}_k$ are UUB, provided the design parameters α, Γ, observer gain L, and control feedback gain K are selected from a specific range of values.

Based on the fact that the estimation error in buffer length and the modeling parameter errors converge to a small compact subset, the attack detectability condition is derived. Upon detecting the attacks, another NN is deployed in order to estimate the flow injected by the attacker. It is shown that the modeling parameter errors of the attack flow also converge to a small compact subset.

Theorem 9.6. *(Condition for attack detectability [16]) Consider the flow model (9.3) and the observer (9.50). Attacks can be detected if the injected (dropped) flow input w_k satisfies*

$$\left\| \sum_{i=0}^{k-1} L^{k-i-1} w_k \right\|$$

$$> \Upsilon_1 + \left\| \sum_{i=0}^{k-1} L^{k-i-1} \left(\tilde{\theta}_k^T \varphi(x_k) + \varepsilon_k + d_k \right) \right\|, \qquad (9.53)$$

where Υ_1 is the threshold for attack detection which is derived from the system behavior in the healthy condition [16].

Remark 9.8. This detectability condition is a theoretical condition by which a class of attacks can be detected. However, this is not the way how attack detection is done in practice. Instead, the network detection residual is constantly monitored and the attack is detected once the residual exceeds the bound Υ_1.

Upon detecting the attack given in terms of bounded traffic flow input, this theorem also demonstrates that the buffer length estimation error and the parameter estimation error are bounded.

Theorem 9.7. *(Estimation of attack input [16]) Consider the flow model at the bottleneck node given as (9.3) and the NN observer described by (9.50). Assume that the attack flow can be modeled as* $w_k = \theta_{w,k}^T \varphi_w(x_k) + \varepsilon_{w,k}$, *where* $\theta_{w,k}$, $\varphi_w(\cdot)$, *and* $\varepsilon_{w,k}$ *are the constant but unknown target weight vector, activation function, and the modeling error of the attack flow, respectively. Then attacks can be detected when the network detection residual exceeds a predefined threshold given by* Υ_1 *in Theorem 9.6. Upon detection, apply the following modified observer:*

$$\hat{x}_{k+1} = \hat{\theta}_k^T \varphi(x_k) + T u_k - L_w(x_k - \hat{x}_k) + \hat{\theta}_{w,k}^T \varphi_w(x_k),$$
$$(9.54)$$

where L_w *is the observer gain matrix and* $\hat{\theta}_k$ *is the actual weight vector for estimation of the unknown nonlinear function* f, *as updated by*

$$\hat{\theta}_{k+1} = \hat{\theta}_k + \alpha_1 \varphi(x_k) \tilde{x}_{k+1}^T$$
$$- \Gamma_1 \left\| I - \alpha_1 \varphi(x_k) \varphi^T(x_k) \right\| \hat{\theta}_k. \quad (9.55)$$

Similarly, $\hat{\theta}_{w,k}$ *is the actual weight vector for estimation of the attack flow and it is tuned using*

$$\hat{\theta}_{w,k+1} = \hat{\theta}_{w,k} + \alpha_2 \varphi_w(x_k) \tilde{x}_{k+1}^T$$
$$- \Gamma_2 \left\| I - \alpha_2 \varphi_w(x_k) \varphi_w^T(x_k) \right\| \hat{\theta}_{w,k}, \quad (9.56)$$

where $\alpha_1, \alpha_2, \Gamma_1, \Gamma_2 \in \mathbb{R}^+$ *are the constant design parameters. Then the buffer length estimation error* $\tilde{x}_k$, *the parameter error* $\tilde{\theta}_k$ *for estimation of* f, *and the parameter* $\hat{\theta}_{w,k}$ *for estimation of attack flow are UUB, if the value of the design parameters are selected from a specific range as derived in [16].*

9.4.2 Attack Detection and Estimation in Nonlinear Physical Systems

Let us consider the following stochastic continuous-time nonlinear system dynamics:

$$\dot{x}_p(t) = f_p(x_p(t)) + \gamma(t)g(x_p(t))u_p(t - \tau(t)) + w_p(x_p(t)), \quad (9.57)$$

where $x_p(t) \in \mathbb{R}^n$, $u_p(t) \in \mathbb{R}^m$, and $w_p(t) \in \mathbb{R}^n$ are the system states, control input, and attack input, respectively. The subscript p is used as a notion of the physical system and is utilized to differentiate from the variables associated with the network; $\tau(t)$ is the time-varying delay and $\gamma(t) \in \mathbb{R}^{n \times n}$ is the packet loss indicator which equals to the identity matrix when the packet is received and to the zero matrix when the packet is lost. Define the bound for time-varying delays as $\tau(t) \leq \bar{d}T$ with $\bar{d}$ as a nonnegative integer and T as the fixed sampling interval.

Remark 9.9. The term $w_p(t)$ is used to characterize the change in system states caused by attacks on the sensors or actuators. Attacks on the physical systems can be detected if $w_p(t)$ satisfies a certain condition. This will be discussed later in this section, after the healthy case, i.e., $w_p(t) = 0$.

Assumption. Let all the assumptions presented in [19] hold. □

After incorporating the time-varying delays and packet losses into the dynamics, the discretized version of the continuous-time system (9.57) is given by

$$z_{k+1} = F(z_k) + G(z_k)u_{p,k} + W_{p,k}, \quad (9.58)$$

where $z_k = [\ x_{p,k}^T \quad u_{p,k-1}^T \quad \cdots \quad u_{p,k-\bar{d}}^T\ ]^T \in \mathbb{R}^{n+\bar{d}m}$ is the augmented state vector of the discretized system, $F(z_k) \in \mathbb{R}^{(\bar{d}m+n) \times 1}$ and $G(z_k) \in \mathbb{R}^{(\bar{d}m+n) \times m}$ are the discretized system matrices defined in [27], and $W_{p,k} \in \mathbb{R}^{(\bar{d}m+n) \times 1}$ is the discretized attack input vector.

The triggering instants are defined as a subsequence $\{k_i\}_{i=1}^\infty$ of $k \in \mathbb{N}$ with initial time instant being $k_0 = 0$. Let $\check{x}_k$ be the states held at the controller, given as

$$\check{x}_k = x_{p,k_i}, \quad \text{for } k_i \leq k < k_{i+1}, \quad (9.59)$$

with the initial state being $\check{x}_0 = x_0$. Then the augmented event sampled state vector becomes

$$\check{z}_k = z_{k_i}, \quad k_i \leq k < k_{i+1},$$

where $\check{z}_k = \begin{bmatrix} \check{x}_k^T & u_{p,k-1}^T \cdots u_{p,k-\bar{d}}^T \end{bmatrix}^T$ and $z_{k_i} = \begin{bmatrix} x_{p,k_i}^T & u_{p,k_i-1}^T \cdots u_{p,k_i-\bar{d}}^T \end{bmatrix}^T$.

The event trigger error is given as

$$\underset{\tau,\gamma}{E}\{e_{ET,k}\} = \underset{\tau,\gamma}{E}\{z_k\} - \underset{\tau,\gamma}{E}\{\check{z}_k\}, \quad k_i \leq k < k_{i+1}. \quad (9.60)$$

The following stochastic value function in terms of augmented state vector is considered for the optimal control design:

$$V_k = \mathop{E}_{\tau,\gamma}\left\{\sum_{j=k}^{\infty} z_j^T Q_z z_j + u_{p,j}^T R_z u_{p,j}\right\},$$
$$k = 0, 1, 2, \cdots, \tag{9.61}$$

where $Q_z \in \mathbb{R}^{(\bar{d}m+n)\times(\bar{d}m+n)} \geq 0$ and $R_z \in \mathbb{R}^{m\times m} > 0$ are the penalty matrices. The optimal control input $u_{p,k}^*$ is then computed as

$$u_{p,k}^* = -\frac{1}{2}R_z^{-1}G^T(\breve{z}_k)\frac{\partial V_{k+1}^*}{\partial z_{k+1}}, \quad k_i \leq k < k_{i+1},$$

where V_{k+1}^* is the optimal stochastic value function in (9.61). However, the optimal control input is usually difficult to obtain because (1) it is very challenging to solve the Hamilton–Jacobi–Bellman (HJB) equation in discrete-time and (2) the nonlinear matrix function $G(z_k)$ is considered unknown. Therefore, an NN-based adaptive solution is followed.

For the design of the control algorithm, initially we assume that there is no attack in the system, i.e., $W_{p,k} = 0$. The dynamics of the system (9.58) in the absence of an attack can be written as

$$z_{k+1} = \begin{bmatrix} F(z_k) & G(z_k) \end{bmatrix}\begin{bmatrix} 1 & u_{p,k}^T \end{bmatrix}^T$$
$$\triangleq \mathop{E}_{\tau,\gamma}\{\theta_I^T \varphi_I(\breve{z}_k)\bar{u}_k + \bar{\varepsilon}_{e,I}(\breve{z}_k, e_{ET,k})\}, \tag{9.62}$$

where $\theta_I = \begin{bmatrix} \theta_F^T & \theta_G^T \end{bmatrix}^T \in \mathbb{R}^{(m+1)l_I \times(\bar{d}m+n)}$ is the constant but unknown target NN weight vector with $\theta_F \in \mathbb{R}^{l_I \times(\bar{d}m+n)}$ and $\theta_G \in \mathbb{R}^{ml_I \times(\bar{d}m+n)}$ are the target weights for the estimation of functions F and G, respectively; $\varphi_I(\breve{z}_k)$ is the activation function selected as $\varphi_I(\breve{z}_k) = diag\{\varphi_F(\breve{z}_k) \quad \varphi_G(\breve{z}_k)\}$, where $\varphi_F(\breve{z}_k) \in \mathbb{R}^{l_I}$, $\varphi_G(\breve{z}_k) \in \mathbb{R}^{ml_I \times m}$ are the activation functions for estimation of F and G, respectively, with l_I being the number of neurons in the hidden layer. The vector $\bar{u}_k = \begin{bmatrix} 1 & u_{p,k}^T \end{bmatrix}^T$ is the augmented control input vector and $\bar{\varepsilon}_{e,I}(\breve{z}_k, e_{ET,k}) = \theta_I^T[\varphi_I(\breve{z}_k + e_{ET,k}) - \varphi_I(\breve{z}_k)]\bar{u}_k + \bar{\varepsilon}_I(\breve{z}_k + e_{ET,k})$ is the event sampled reconstruction error where $\bar{\varepsilon}_I(\breve{z}_k + e_{ET,k}) = \varepsilon_I\bar{u}_k$ and $\varepsilon_I = \begin{bmatrix} \varepsilon_F(z_k) & \varepsilon_G(z_k) \end{bmatrix}$ is the NN reconstruction error, with $\varepsilon_F(z_k)$ and $\varepsilon_G(z_k)$ being the NN reconstruction error for estimation of the functions F and G, respectively.

Let the event-based identifier dynamics be defined as

$$\hat{z}_{k+1} = \hat{F}(\breve{z}_k) + \hat{G}(\breve{z}_k)u_{p,k} \triangleq \mathop{E}_{\tau,\gamma}\{\hat{\theta}_I^T \varphi_I(\breve{z}_k)\bar{u}_k\}, \tag{9.63}$$

with $\hat{\theta}_I$ being the actual NN weights of the identifier. Define the estimation error of the identifier as $\mathop{E}_{\tau,\gamma}\{\tilde{z}_k\} = \mathop{E}_{\tau,\gamma}\{z_k\} - \mathop{E}_{\tau,\gamma}\{\hat{z}_k\}$. Then the identification error dynamics is found as

$$\mathop{E}_{\tau,\gamma}\{\tilde{z}_{k+1}\} = \mathop{E}_{\tau,\gamma}\left\{\begin{array}{c}\tilde{\theta}_{I,k}^T\varphi_I(z_k)\bar{u}_k + \hat{\theta}_I^T[\varphi_I(z_k) \\ -\varphi_I(\breve{z}_k)]\bar{u}_k + \bar{\varepsilon}_I(z_k)\end{array}\right\}, \tag{9.64}$$

where $\tilde{\theta}_{I,k} = \theta_{I,k} - \hat{\theta}_{I,k}$ is the identifier NN weight estimation error. The NN weights update rule for the identifier is selected as

$$\mathop{E}_{\tau,\gamma}\{\hat{\theta}_{I,k}\}$$
$$= \mathop{E}_{\tau,\gamma}\left\{\hat{\theta}_{I,k-1} + \frac{\chi_k\alpha_I\varphi_I(\breve{z}_{k-1})\bar{u}_{k-1}\tilde{z}_k^T}{(\varphi_I(\breve{z}_{k-1})\bar{u}_{k-1})^T(\varphi_I(\breve{z}_{k-1})\bar{u}_{k-1})+1}\right\}, \tag{9.65}$$

with $\alpha_I > 0$ being the learning rate and χ_k being the event indicator which equals to one in case an event is triggered and zero otherwise.

To find the optimal control input, an event-triggered stochastic ADP using the actor-critic networks is used where the solution to the discrete HJB is approximated by the critic network and the optimal control input is approximated by the actor network. The value function in (9.61) is estimated by the critic NN as

$$\hat{V}_k = \mathop{E}_{\tau,\gamma}\{\hat{\theta}_{V,k}^T\varphi_V(\breve{z}_k)\}, \tag{9.66}$$

where $\hat{V}_k$ is the estimated value function at time instant k, $\theta_{V,k}$ is the actual critic NN weight vector at time instant k, and φ_V denotes the activation function for the critic NN. The weight update law for the critic network is selected as

$$\mathop{E}_{\tau,\gamma}\{\hat{\theta}_{V,k}\} = \mathop{E}_{\tau,\gamma}\left\{\hat{\theta}_{V,k-1} - \frac{\chi_k\alpha_V \Delta\varphi_V(\breve{z}_k)e_{V,k}}{\Delta\varphi_V^T(\breve{z}_k)\Delta\varphi_V(\breve{z}_k)+1}\right\}, \tag{9.67}$$

where $\alpha_V > 0$ is the learning rate, $e_{V,k}$ is the value function approximation error defined as $e_{V,k} = V_k - \hat{V}_k$, and $\Delta\varphi_V(\breve{z}_k)$ is defined as $\Delta\varphi_V(\breve{z}_k) = \varphi_V(\breve{z}_{k+1}) - \varphi_V(\breve{z}_k)$.

The third NN or the actor NN is used to approximate the optimal control input, as given by

$$u_{p,k} = \mathop{E}_{\tau,\gamma}\{\hat{\theta}_{u,k}^T\varphi_u(\breve{z}_k)\}, \tag{9.68}$$

where $\theta_{u,k}$ is the actual actor NN weight vector at time instant k and φ_u denotes the activation function for the actor NN. The weight update law for the actor network is selected as

$$\underset{\tau,\gamma}{E}\{\hat{\theta}_{u,k}\} = \underset{\tau,\gamma}{E}\left\{\hat{\theta}_{u,k-1} - \frac{\chi_k \alpha_u \varphi_u(\check{z}_{k-1})e_{u,k-1}^T}{\varphi_u^T(\check{z}_{k-1})\varphi_u(\check{z}_{k-1}) + 1}\right\},$$

$$(9.69)$$

where $\alpha_u > 0$ is the learning rate and $e_{u,k}$ is the control input estimation error at time instant k defined as

$$\underset{\tau,\gamma}{E}\{e_{u,k}\} = \underset{\tau,\gamma}{E}\left\{\hat{\theta}_{u,k}^T \varphi_u(\check{z}_k)\right.$$

$$\left. + \frac{1}{2}R_z^{-1}\hat{G}^T(\check{z}_k)\left(\frac{\partial \varphi_V(\check{z}_{k+1})}{\partial z_{k+1}}\right)\hat{\theta}_{V,k}\right\},$$

$$k_i \leq k < k_{i+1}. \qquad (9.70)$$

The event triggering condition design is critical because on the one hand, excessive triggering clearly deviates from the original intention of reducing the data transmission. On the other hand, insufficient triggering will result in a regulation error, thus degrading the performance and even leading to instability of the system. Here, the event triggering condition is designed in an adaptive fashion [19], as given by

$$\sigma_{ET,k} D\left(\left\|\underset{\tau,\gamma}{E}\{e_{ET,k}\}\right\|^2\right) > \rho_p \left\|\underset{\tau,\gamma}{E}\{z_k\}\right\|^2, \quad (9.71)$$

where the time-varying function $\sigma_{ET,k}$ and the constant ρ_p are defined in [16] and $D(\cdot)$ denotes the dead-zone operator, which is defined as the function $D(x) = 0$ when $\left\|\underset{\tau,\gamma}{E}\{z_k\}\right\| > B_z$ and as $D(x) = x$ otherwise, where B_z is the upper bound of the system states. Now it is shown that the system states are bounded in the healthy case in the absence of attacks on the network and the physical system.

It was shown in [19] that in the absence of attacks on the network and the physical system, the states $\underset{\tau,\gamma}{E}\{z_k\}$, identifier NN estimation error $\underset{\tau,\gamma}{E}\{\tilde{\theta}_{I,k}\}$, critic NN estimation error $\underset{\tau,\gamma}{E}\{\tilde{\theta}_{V,k}\}$, and actor NN estimation error $\underset{\tau,\gamma}{E}\{\tilde{\theta}_{u,k}\}$ are UUB in the mean, when the event triggering condition is chosen as (9.71), update laws of the NNs are chosen as (9.65), (9.67), (9.69) for the identifier, critic, and actor, respectively, and the learning rate constants $\alpha_I, \alpha_V, \alpha_u$ are chosen from specific bounds.

So far it was assumed that the communication network is in the healthy condition, i.e., the delays and packet losses are upper bounded by small values. Now it is important to determine the maximum values of these delays and losses that the system can tolerate without becoming unstable.

Theorem 9.8. *(System stability under network attacks [16]) Consider the nonlinear physical system dynamics (9.58) without attacks on the physical system, i.e., $W_{p,k} = 0$. Assume that the communication network is under attack such that $\varepsilon_{p,k}$ is always greater than ε_m, which is defined as the maximum delay and packet loss bound that the system can tolerate without becoming unstable. Then the physical system becomes unstable if ε_m satisfies the following condition:*

$$\eta_k \underset{\tau,\gamma}{E}\left\{\|z_{\varepsilon_m+k_i} - z_{k_i}\|^2\right\} > \underset{\tau,\gamma}{E}\left\{HZ_k - B_{\hat{\theta},k}^{c2}\right\}, \quad (9.72)$$

where the time-varying functions η_k, Z_k, $B_{\hat{\theta},k}^{c2}$ and the constant parameter H are defined in [16].

Stability of the physical system cannot be guaranteed if $\epsilon_{p,k} \geq \epsilon_m$ always holds.

Now we consider attacks on the physical system components. In the next theorem, a detectability condition for such attacks is derived by using the identifier NN. The identifier NN was originally designed to estimate the unknown nonlinear system dynamics, but as we have made the assumption that the attack sets in after the convergence of the NN weights, it can now be utilized to capture the abnormality in the system caused by the attack.

Theorem 9.9. *(Attack detectability condition at the physical system [16]) Consider the nonlinear physical system dynamics (9.62). Let the identifier NN be defined as (9.63) with the weight update law given in (9.65). Then attacks $W_{p,k}$ on the physical system can be detected if $W_{p,k}$ satisfies*

$$\underset{\tau,\gamma}{E}\left\{\left\|\tilde{\theta}_{I,k}^T \varphi_I(z_k)\bar{u}_k + \hat{\theta}_I^T[\varphi_I(z_k) - \varphi_I(\check{z}_k)]\bar{u}_k + \bar{\varepsilon}_I(z_k)\right.\right.$$

$$\left.\left. + W_{p,k}\right\|\right\} > B_{z,k}^{c2},$$

$$(9.73)$$

where $B_{z,k}^{c2}$ is the attack detection threshold and is defined in [16].

Next, upon detecting the attack on the physical systems, it is of interest to estimate the attacker input $W_{p,k}$.

In order to do this, we rewrite the physical system dynamics (9.62) as

$$z_{k+1} = \begin{bmatrix} F(z_k) & W_{p,k} & G(z_k) \end{bmatrix} \begin{bmatrix} 1 & 1 & u_{p,k}^T \end{bmatrix}^T$$
$$= \underset{\tau,\gamma}{E} \{\theta_{Iw}^T \varphi_I(\check{z}_k) \bar{u}_{w,k} + \bar{\varepsilon}_{e,Iw}(\check{z}_k, e_{ET,k})\}, \quad (9.74)$$

where $\theta_{Iw} = \begin{bmatrix} \theta_F^T & \theta_W^T & \theta_G^T \end{bmatrix}^T$ is the constant but unknown target NN weight vector and $\bar{u}_{w,k} = \begin{bmatrix} 1 & 1 & u_k^T \end{bmatrix}^T$ is the augmented control input vector. The event sampled reconstruction error $\bar{\varepsilon}_{e,Iw}(\check{z}_k, e_{ET,k})$ is then given by

$$\bar{\varepsilon}_{e,Iw}(\check{z}_k, e_{ET,k}) = \theta_{Iw}^T \left(\varphi_I(\check{z}_k + e_{ET,k}) - \varphi_I(\check{z}_k) \right) \bar{u}_{w,k}$$
$$+ \bar{\varepsilon}_{Iw}(\check{z}_k + e_{ET,k})$$
$$= \theta_{Iw}^T \left(\varphi_I(\check{z}_k + e_{ET,k}) - \varphi_I(\check{z}_k) \right) \bar{u}_{w,k}$$
$$+ \varepsilon_{Iw} \bar{u}_{w,k}, \quad (9.75)$$

with $\bar{\varepsilon}_{Iw} = \begin{bmatrix} \varepsilon_F(z_k) & \varepsilon_W(z_k) & \varepsilon_G(z_k) \end{bmatrix}$ being the NN reconstruction error vector.

Accordingly, an NN-based approximator is added to modify the identifier (9.63) for the purpose of estimating the attack signal upon detection. The dynamics of the modified NN identifier become

$$\hat{z}_{k+1} = \hat{F}(\check{z}_k) + \hat{W}_{p,k} + \hat{G}(\check{z}_k) u_{p,k}$$
$$= \underset{\tau,\gamma}{E} \{\hat{\theta}_{Iw}^T \varphi_I(\check{z}_k) \bar{u}_{w,k}\}, \quad (9.76)$$

where $\hat{\theta}_{Iw}$ are the actual NN weights of the identifier, $\hat{z}_{k+1}$ denotes the estimated augmented state at time instant $k + 1$, and $\hat{F}, \hat{G}, \hat{W}_{p,k}$ are the estimates of the nonlinear functions F, G and the attack input $W_{p,k}$, respectively. Then the error dynamics of the identifier can be computed as

$$\underset{\tau,\gamma}{E} \{\tilde{z}_{k+1}\} = \underset{\tau,\gamma}{E} \left\{ \tilde{\theta}_{Iw,k}^T \varphi_I(z_k) \bar{u}_{w,k} + \bar{\varepsilon}_{Iw}(z_k) \right.$$
$$\left. + \hat{\theta}_{Iw}^T [\varphi_I(z_k) - \varphi_I(\check{z}_k)] \bar{u}_{w,k}, \right\}, \quad (9.77)$$

where $\tilde{\theta}_{Iw,k} \triangleq \theta_{Iw,k} - \hat{\theta}_{Iw,k}$ is the identifier NN weight estimation error.

Upon detection, the next theorem presents the update law of the attack estimation NN identifier.

Theorem 9.10. *(Identifier-based attack estimation [16])*
Upon detecting the attack, consider the identifier NN given

by (9.76) with the following weight update law:

$$\underset{\tau,\gamma}{E} \{\hat{\theta}_{Iw,k}\}$$
$$= \underset{\tau,\gamma}{E} \left\{ \hat{\theta}_{Iw,k-1} + \frac{\chi_k \alpha_{Iw} \varphi_I(\check{z}_{k-1}) \bar{u}_{w,k-1} \check{z}_k^T}{(\varphi_I(\check{z}_{k-1}) \bar{u}_{w,k-1})^T (\varphi_I(\check{z}_{k-1}) \bar{u}_{w,k-1}) + 1} \right\}, \quad (9.78)$$

where $\alpha_{Iw} \in \mathbb{R}^+$ is the learning rate, which is to be chosen as $0 < \alpha_{Iw} < 1/2$, and $\underset{\tau,\gamma}{E} \{\check{z}_k\} = \underset{\tau,\gamma}{E} \{z_k\} - \underset{\tau,\gamma}{E} \{\hat{z}_k\}$ is the identifier NN estimation error. Then, the identifier NN weight estimation error $\underset{\tau,\gamma}{E} \left(\tilde{\theta}_{Iw,k} \right)$ is UUB in the mean.

9.5 RESULTS AND DISCUSSION

Simulation results for the algorithms discussed in the previous sections are presented here. All the simulations are performed in Matlab.

9.5.1 Simulation Results for Linear NCS

For the network side, the following parameters were used to simulate the algorithms: sampling period $T = 100$ ms, total simulation time $T_s = 500\,T$, standard transmission rate $v_0 = 300$ packets per T, the desired flow in the bottleneck node $\rho_0 = 300$ packets, $m = 3$, $l_1 = 1/8$, $l_1 = 1/4$, $l_1 = 1/2$, and P and R are identity matrices.

Scenario A1 (Normal Case)
Fig. 9.2A shows that in the absence of attacks, the Q-function estimator (QFE) error becomes very close to zero. Moreover, it shows that the number of packets in the bottleneck node fluctuates around the desired value.

Scenario A2 (Network Under Attack)
In this case, a jamming attack is introduced at $t = 250T$. As depicted in Fig. 9.2B, the attacker increases the number of jammers in the network linearly with time until the maximum value. As a result, the packets injected by the attacker will increase to the maximum of 500 packets per sampling period. The estimation error of the flow exceeds the threshold shortly after the attack is launched and thus, it can be detected. Upon detection, the observer for attack estimation is applied and the estimated attack flow is shown in Fig. 9.2B.

For the simulations on the physical system side, the benchmark example of batch reactor [28] is considered whose continuous-time dynamics are given as

$$\dot{x}(t) = A_p x(t) + B_p u(t), \quad (9.79)$$

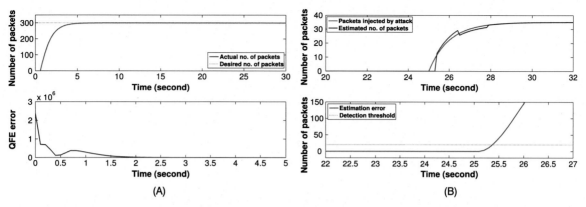

FIG. 9.2 (A) A1: Actual and desired number of packets in the bottleneck node, QFE error. **(B)** A2: Actual and estimated attack input, estimation error, and detection threshold.

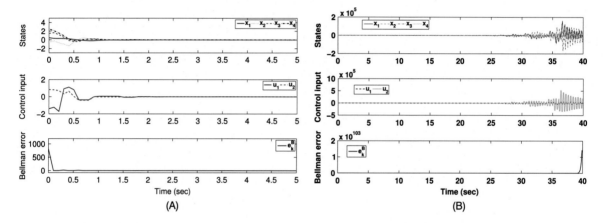

FIG. 9.3 System states, input, and Bellman error in **(A)** Scenario B1, **(B)** Scenario B2.

where

$$A_p = \begin{bmatrix} 1.38 & -0.2077 & 6.715 & -5.676 \\ -0.5814 & -4.29 & 0 & 0.675 \\ 1.067 & 4.273 & -6.654 & 5.893 \\ 0.048 & 4.273 & 1.343 & -2.104 \end{bmatrix},$$

$$B_p = \begin{bmatrix} 0 & 0 \\ 5.679 & 0 \\ 1.136 & -3.146 \\ 1.136 & 0 \end{bmatrix}.$$

The system is discretized with sampling time $T = 0.1$ sec and a total simulation time of 50 sec. At first, performance of the optimal event-triggered controller in the healthy condition is evaluated. Then it is shown that the system becomes unstable when the delays and packet losses exceed a limit due to network attacks. Finally it is demonstrated that the proposed scheme is

able to detect attacks on the physical system. The initial state of the system is selected as $\begin{bmatrix} 2 & -2 & 6 & 10 \end{bmatrix}^T$. A quadratic cost function is selected with penalty matrices $H_z = diag\{0.2I_{4\times4}, 0.1I_{2\times2}, 0.1I_{2\times2}\}$, $R_z = diag\{0.1I_{2\times2}, 0.1I_{2\times2}\}$, $\gamma_z = 5$, where $I_{n\times n}$ denotes the $n \times n$ identity matrix. The learning gains are $\alpha_g = 0.5$, $\alpha_w = 1$. Initial value of the gain matrix is $W_0 = 9999I$. To ensure that the initial value of the Q-function parameters are nonzero and in a compact set, the values are selected as 60% of the actual values. A noise of small magnitude, generated randomly from a normal distribution in the range [0, 0.5], is added to the regression vector to satisfy the PE condition. A Monte Carlo simulation was run for all the scenarios considered.

Scenario B1 (Normal Case)

The network is assumed to be in the healthy condition. The delay bound is two sampling instants, i.e., $d = 2$,

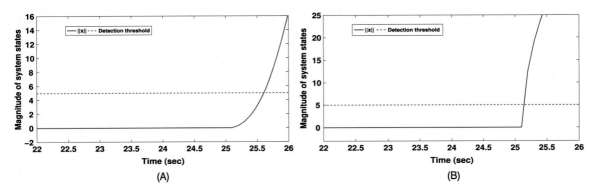

FIG. 9.4 Magnitude of system states under **(A)** actuator attack **(B)** sensor attack.

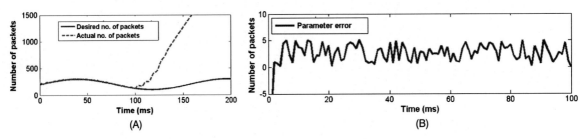

FIG. 9.5 **(A)** Actual and desired number of packets in bottleneck node. **(B)** Parameter error before the attack is launched.

with a mean delay of 0.1 sec, and the packet loss follows a Bernoulli distribution with the probability of dropping packets being 0.1. As plotted in Fig. 9.3A, the system states converge to zero after a few seconds, and the control input also reaches zero as the system states converge. The convergence of the Bellman error verifies that the control input reaches optimality.

Scenario B2 (Network Under Attack)
In this scenario, it is supposed that the network is under attack such that the maximum delay is 0.25 sec and the packet loss rate is 0.2. As a result, the overall delay exceeds the maximum value that the system can tolerate. As depicted in Fig. 9.3B, the system states, control input, and Bellman error diverge, which implies that the system becomes unstable.

Scenario B3 (Physical System Under Attack)
In this scenario, actuator and sensor attacks on the physical system are injected and the attack detectability of the controlled system is verified. The detection threshold is chosen as 5. First an actuator attack in introduced at $t_a = 250T$, where the input is manipulated from $u_{p,k}$ to $u_{p,k} + \Delta u_{p,k}$ with $\Delta u_{p,k} = 1.2(k - 250)$. The magnitude of system states starts increasing and crosses the

detection threshold soon after the launch of the attack. Thus the attack can be detected, as shown in Fig. 9.4A.

Next, a sensor attack is introduced where the measurement is manipulated from $x_{p,k}$ to $x_{p,k} + \Delta x_{p,k}$, with $\Delta x_{p,k} = \begin{bmatrix} 20 & 20 & 20 & 20 \end{bmatrix}^T$. In this case too, $\|x\|$ goes unbounded, as shown in Fig. 9.4B, crossing the threshold after the launch of the attack, thus the attack can be detected.

9.5.2 Simulation Results for Nonlinear NCS
For the communication network simulation, the sampling period is chosen as $T = 1$ ms, and the total simulation time is $T_s = 200T$. Assume the desired number of packets in the bottleneck node to be expressed as $x_{d,k} = 200 + 100 \sin(k/25)$, the unknown nonlinear dynamics to be $f(x_k) = \sqrt{x_k}$, and the maximum modeling error or disturbance to be $d_M = 5$. A single-layer NN consisting of five neurons is considered as the observer, and the initial values of the NN weights are selected as zero. The basis function is chosen to be a polynomial of the states, i.e., $\phi(x_k) = \begin{bmatrix} 1 & x_k & x_k^2 & x_k^3 & x_k^4 \end{bmatrix}$. The feedback control gain $K = 0.05$, the observer gain $L = 0.1$, and the coefficient of the adaptive term $\Gamma = 0.5$. The initial adaptation gain α is chosen as 0.1 and

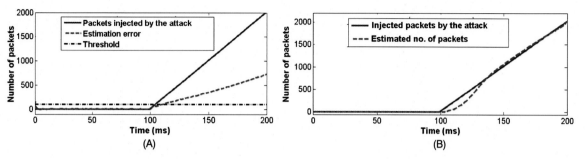

FIG. 9.6 **(A)** Estimation error and attack injected packets. **(B)** Actual and estimated number of packets by the attacker.

is updated using the projection algorithm as given by $\alpha_k = 0.5 / \left(0.1 + \varphi^T (x_k) \varphi (x_k) \right)$. Fig. 9.5A shows the desired and actual number of packets in the bottleneck node. Before the launch of the attack, the actual number of packets is close to the desired value. Moreover, as shown in Fig. 9.5B, the modeling parameter error is close to zero before the attack is launched. A jamming attack is introduced at $t = 100$ ms. Assuming the attacking strength increases linearly, the number of injected packets is modeled as $\omega_k = \alpha_w (k - k_0)$, where α_w is the attack strength and k_0 is the attack launch time. Here $\alpha_w = 20$ and $k_0 = 100$. Fig. 9.6A shows the estimation error and the attacker-injected packets, with the adaptive observer (9.50) incorporated. Before the launch of the attack, the estimation error stays close to zero, which means that the state observer presented in (9.50) is fairly accurate. Once the attack is launched, the number of packets in the bottleneck node starts increasing and deviates from the desired value, as shown in Fig. 9.5A. As a result, the estimation error, plotted in Fig. 9.6A, exceeds the threshold shortly after the attack is launched and thus the attack can be detected. The threshold for the detection of attack is selected as 50.

Upon detection, the observer for attack estimation with gain $L_w = 0.1$ is applied to estimate the number of packets injected by the attacker. The observer is formulated as a one-layer NN having four neurons in the hidden layer. The basis function φ_w is chosen to be the same as that in the flow observer. The observer gain $\Gamma_2 = 0.001$. The adaptation gain α_w is updated using the same projection algorithm as α, with an initial value of 0.025. As plotted in Fig. 9.6B, the estimated number of packets injected by the attacker with the new observer converges to the actual value. With the estimated attack flow, one can estimate the delay and packet losses in the link, which can be further utilized to tune the controller parameters of the physical systems.

For the physical system simulation, the following second-order nonlinear system dynamics [20] is considered:

$$
\begin{aligned}
x_{1,k+1} &= x_{2,k}, \\
x_{2,k+1} &= \frac{x_{2,k}}{1 + x_{1,k}^2} + \left(2 + \sin\left(x_{1,k}\right)\right) u_k.
\end{aligned}
\tag{9.80}
$$

The sampling period T_s is chosen as 0.01 seconds and the total simulation time is $1500 T_s$. The time-varying delay bound is $\bar{d} = 2$, the mean value of delay is $\underset{\tau,\gamma}{E} (\tau) = 12$ ms. The packet losses follow a Bernoulli distribution with the probability of dropping packets being $p = 0.1$. The penalty matrices are chosen to be $P_z = 0.1 * I_{2 \times 2}$ and $R_z = 1$. The initial states are selected as $[-2, 2]^T$. The number of neurons for the identifier is 50, for the actor and critic the numbers are 15 and 39, respectively, with all the weights initially assigned at random from a uniform distribution in the range $[0, 1]$. The activation functions of the NN are selected as polynomials. The critic activation function is selected as $\varphi_V = tanh(\breve{z}_{1,k}^2; \breve{z}_{1,k} \breve{z}_{2,k}; \cdots; \breve{z}_{2,k}^2; \breve{z}_{2,k} \breve{z}_{3,k}; \cdots; \breve{z}_{4,k}^2; \breve{z}_{1,k}^4; \breve{z}_{1,k}^3 \breve{z}_{2,k}; \cdots; \breve{z}_{1,k} \breve{z}_{2,k} \breve{z}_{3,k} \breve{z}_{4,k}; \breve{z}_{4,k}^4) \in \mathbb{R}^{39}$, and the actor NN and identifier NN activation functions are $\varphi_u(\cdot) = tanh(1; \breve{z}_{1,k}; \breve{z}_{2,k}; \breve{z}_{3,k}; \breve{z}_{4,k}; \breve{z}_{1,k}^2; \breve{z}_{1,k} \breve{z}_{2,k}; \cdots; \breve{z}_{2,k}^2; \breve{z}_{2,k} \breve{z}_{3,k}; \cdots; \breve{z}_{4,k}^2;) \in \mathbb{R}^{15}$ and $\varphi_I(\cdot) = tanh(1; \breve{z}_{1,k}^3; \breve{z}_{1,k}^2 \breve{z}_{2,k}; \cdots; \breve{z}_{1,k} \breve{z}_{3,k} \breve{z}_{4,k}; \breve{z}_{2,k}^3; \cdots \breve{z}_{4,k}^3; \breve{z}_{1,k}^4; \breve{z}_{1,k}^3 \breve{z}_{2,k}; \cdots; \breve{z}_{1,k} \breve{z}_{2,k} \breve{z}_{3,k} \breve{z}_{4,k}; \breve{z}_{4,k}^4) \in \mathbb{R}^{50}$, respectively. The learning gain values are selected as follows: $\alpha_V = 0.01$, $\alpha_u = 0.004$, and $\alpha_I = 0.001$. The parameters for the event triggering condition are $K^* = 0.45$, $\Gamma = 0.99$, $G_M = 3.5$, and the Lipschitz constant for the identifier and actor NN are computed to be 6.5 and 5.2, respectively.

First the healthy scenario, where there is no attack on the physical system or the communication links, is considered. As plotted in Fig. 9.7A, the system states

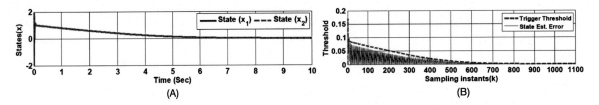

FIG. 9.7 **(A)** Convergence of the states when there is no attack. **(B)** Trigger threshold and state estimation error.

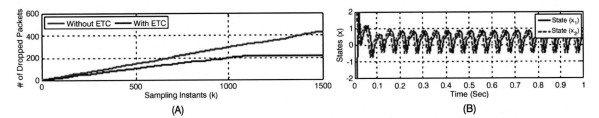

FIG. 9.8 **(A)** Number of dropped packets with and without event-triggered control. **(B)** System states when overall delay exceeds the threshold.

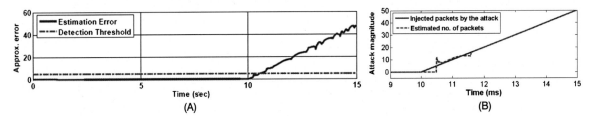

FIG. 9.9 Attack on physical system. **(A)** Estimation error and detection threshold. **(B)** Actual and estimated attack input.

converge to values close to zero after about 7 seconds, although the initial values of system states and NN weights are fairly far from their target values. Fig. 9.7B shows the evolution of the triggering threshold and the state estimation error. The state estimation error oscillates between zero and the trigger threshold due to the fact that in event trigger control methods, the estimation error is set to zero when it becomes equal or greater than the trigger threshold. Also the state estimation error converges to values close to zero after about 10 seconds and eventually stops satisfying the trigger condition due to the dead-zone function.

Next, in order to verify that the ETC scheme helps reduce the network packet losses, a black hole attack is introduced to the network. To be specific, we assume at each sampling instant the attack drops the sensor-to-controller packet with the probability of 0.3. Fig. 9.8A shows the comparison of the accumulated number of dropped packets between the event-triggered and time-driven control systems in the presence of a black hole

attack on the network. It can be observed that for the ETC, the number of dropped packets by the attack is much lower than that of the time-driven system. Especially when the event trigger error is small enough (after 11 s in this example) and the event is no longer triggered, there will be no data loss at all. Therefore, it is confirmed that the ETC scheme reduces the packet losses in the presence of attacks.

Next, a jamming attack is introduced in the network due to which the overall delay exceeds the maximum tolerable value. For the simulation, we choose $\epsilon_m = 6T$ such that inequality (9.72) holds. Then, as shown in Fig. 9.8B, the states of the system do not converge.

Now, an attack on the physical system is introduced while keeping the network in the healthy condition. Assume the attack is launched at $t_{at} = 10$ s after the convergence of the system states. The attack targets the system by modifying the sensor and the state x_2 such that

$$x_{2,k+1} = \frac{x_{2,k}}{1 + x_{1,k}^2} + \left(2 + sin(x_{1,k})\right) u_k + \beta_{at}\left(k - t_{at}\right),$$

where β_{at} is the attacking strength, selected as 0.1 in the simulation. As shown in Fig. 9.9A, the estimation error increases after the launch of the attack and exceeds the detection threshold shortly. As a result, the attack can be detected.

After the detection of the attack, the modified observer for attack estimation is applied. This observer is an identifier NN with 50 neurons; initially the weights were randomly assigned from a uniform distribution in [0,1]. Polynomial basis function is chosen for this NN. As shown in Fig. 9.9B, the estimated attack magnitude given by the new observer converges to the actual attack magnitude about one second after the attack is launched.

9.6 CONCLUSIONS

Networked control systems suffer from security threats due to the presence of communication links that are used for transmitting sensor and controller data. This chapter presents novel attack detection and estimation schemes via Q-learning in the case of a linear system and an NN in the case of a nonlinear system by monitoring the abnormality in the state vector of the system. The cyber- and physical systems are considered separately to design the controllers and security-related algorithms, but it is shown that the attacks on one affect the other. Residual-based detection schemes are presented; in this method, unlike other conventional methods, one can identify the presence of abnormalities in the system before the states start deteriorating. Event triggering control methods are implemented and it is demonstrated that such controllers cause less communication than the regular ones, which also facilitates reduced chances of being attacked. The detection and estimation schemes consider generic attack models; however, certain sophisticated attacks cannot be detected by such methods.

REFERENCES

1. E. Altman, T. Basar, Optimal rate control for high speed telecommunication networks, in: Decision and Control, 1995, Proceedings of the 34th IEEE Conference on, vol. 2, IEEE, 1995, pp. 1389–1394.
2. H. Baumann, W. Sandmann, Markovian modeling and security measure analysis for networks under flooding dos attacks, in: Parallel, Distributed and Network-Based Processing (PDP), 2012 20th Euromicro International Conference on, IEEE, 2012, pp. 298–302.
3. P.-K. Huang, X. Lin, C.-C. Wang, A low-complexity congestion control and scheduling algorithm for multihop wireless networks with order-optimal per-flow delay, IEEE/ACM Transactions on Networking (TON) 21 (2) (2013) 495–508.
4. S. Jagannathan, J. Talluri, Predictive congestion control of ATM networks: multiple sources/single buffer scenario, Automatica 38 (5) (2002) 815–820.
5. J. Jin, W.-H. Wang, M. Palaniswami, A simple framework of utility max-min flow control using sliding mode approach, IEEE Communications Letters 13 (5) (2009).
6. P. Lee, A. Clark, L. Bushnell, R. Poovendran, A passivity framework for modeling and mitigating wormhole attacks on networked control systems, IEEE Transactions on Automatic Control 59 (12) (2014) 3224–3237.
7. F.L. Lewis, D. Vrabie, V.L. Syrmos, Optimal Control, John Wiley & Sons, 2012.
8. C.-p. Li, E. Modiano, Receiver-based flow control for networks in overload, IEEE/ACM Transactions on Networking 23 (2) (2015) 616–630.
9. Z. Lu, W. Wang, C. Wang, Modeling, evaluation and detection of jamming attacks in time-critical wireless applications, IEEE Transactions on Mobile Computing 13 (8) (2014) 1746–1759.
10. F. Miao, M. Pajic, G.J. Pappas, Stochastic game approach for replay attack detection, in: Decision and Control (CDC), 2013 IEEE 52nd Annual Conference on, IEEE, 2013, pp. 1854–1859.
11. Y. Mo, B. Sinopoli, Secure control against replay attacks, in: Communication, Control, and Computing, 2009, Allerton 2009, 47th Annual Allerton Conference on, IEEE, 2009, pp. 911–918.
12. Y. Mo, S. Weerakkody, B. Sinopoli, Physical authentication of control systems: designing watermarked control inputs to detect counterfeit sensor outputs, IEEE Control Systems 35 (1) (2015) 93–109.
13. V. Narayanan, S. Jagannathan, Distributed adaptive optimal regulation of uncertain large-scale interconnected systems using hybrid q-learning approach, IET Control Theory & Applications 10 (12) (2016) 1448–1457.
14. H. Niu, A Control Theoretic Approach for Security of Cyber-Physical Systems, Missouri University of Science and Technology, 2016.
15. H. Niu, S. Jagannathan, Optimal defense and control for cyber-physical systems, in: Computational Intelligence, 2015 IEEE Symposium Series on, IEEE, 2015, pp. 634–639.
16. H. Niu, S. Jagannathan, Neural network-based attack detection in nonlinear networked control systems, in: Neural Networks (IJCNN), 2016 International Joint Conference on, IEEE, 2016, pp. 4249–4254.
17. F. Pasqualetti, F. Dörfler, F. Bullo, Attack detection and identification in cyber-physical systems, IEEE Transactions on Automatic Control 58 (11) (2013) 2715–2729.
18. P.N. Raj, P.B. Swadas, DPRAODV: a dynamic learning system against blackhole attack in AODV based MANET, arXiv preprint, arXiv:0909.2371, 2009.
19. A. Sahoo, S. Jagannathan, Stochastic optimal regulation of nonlinear networked control systems by using event-driven adaptive dynamic programming, IEEE Transactions on Cybernetics 47 (2) (2017) 425–438.

20. A. Sahoo, H. Xu, S. Jagannathan, Adaptive neural network-based event-triggered control of single-input single-output nonlinear discrete-time systems, IEEE Transactions on Neural Networks and Learning Systems 27 (1) (2016) 151–164.

21. Avimanyu Sahoo, Vignesh Narayanan, S. Jagannathan, Optimal event-triggered control of uncertain linear networked control systems: a co-design approach, in: Proceedings of the 2017 IEEE Symposium on Adaptive Dynamic Programming and Reinforcement Learning (ADPRL), IEEE, 2017.

22. K. Sallhammar, B.E. Helvik, S.J. Knapskog, Towards a stochastic model for integrated security and dependability evaluation, in: Availability, Reliability and Security, 2006, ARES 2006, The First International Conference on, IEEE, 2006.

23. H. Sandberg, S. Amin, K.H. Johansson, Cyberphysical security in networked control systems: an introduction to the issue, IEEE Control Systems 35 (1) (2015) 20–23.

24. B. Satchidanandan, P.R. Kumar, Dynamic watermarking: active defense of networked cyber-physical systems, Proceedings of the IEEE 105 (2) (2017) 219–240.

25. P. Tague, D. Slater, R. Poovendran, G. Noubir, Linear programming models for jamming attacks on network traffic flows, in: Modeling and Optimization in Mobile, Ad Hoc, and Wireless Networks and Workshops, 2008, WiOPT 2008, 6th International Symposium on, IEEE, 2008, pp. 207–216.

26. A. Teixeira, I. Shames, H. Sandberg, K.H. Johansson, A secure control framework for resource-limited adversaries, Automatica 51 (2015) 135–148.

27. H. Xu, S. Jagannathan, Stochastic optimal controller design for uncertain nonlinear networked control system via neuro-dynamic programming, IEEE Transactions on Neural Networks and Learning Systems 24 (3) (2013) 471–484.

28. H. Xu, S. Jagannathan, F. Lewis, Stochastic optimal design for unknown linear discrete-time system zero-sum games in input–output form under communication constraints, Asian Journal of Control 16 (5) (2014) 1263–1276.

29. H. Xu, S. Jagannathan, F.L. Lewis, Stochastic optimal control of unknown linear networked control system in the presence of random delays and packet losses, Automatica 48 (6) (2012) 1017–1030.

30. Q. Zhao, H. Xu, J. Sarangapani, Finite-horizon near optimal adaptive control of uncertain linear discrete-time systems, Optimal Control Applications and Methods 36 (6) (2015) 853–872.

31. M. Zhu, S. Martínez, On the performance analysis of resilient networked control systems under replay attacks, IEEE Transactions on Automatic Control 59 (3) (2014) 804–808.

32. Q. Zhu, T. Başar, Robust and resilient control design for cyber-physical systems with an application to power systems, in: Decision and Control and European Control Conference (CDC-ECC), 2011 50th IEEE Conference on, IEEE, 2011, pp. 4066–4071.

CHAPTER 10

Sensitivity Analysis With Artificial Neural Networks for Operation of Photovoltaic Systems

O. MAY TZUC, MENG • A. BASSAM, PHD • L.J. RICALDE, PHD • E. CRUZ MAY, MENG

10.1 INTRODUCTION

In terms of energy and sustainability, photovoltaic (PV) systems are one of the most mature technologies for the use of solar resources. They offer a direct conversion of solar energy into electricity without moving parts and without producing atmospheric emissions [1]. On PV modules, the high levels of irradiance favor the generation of electrical energy; however, due to climatic conditions, the radiation at peak sun hours implies a negative impact on the electrical conversion efficiency [5]. For a typical commercial PV module operating at its maximum power point, only 10 to 15% of the incident light is converted into electricity, which means that a large part of the energy is converted into heat. Prolonged exposure to radiation increases the photon flow into the PV cells, resulting in the increase of the PV module temperature to values higher than the environmental temperature [15]. This causes the band gap of the PV semiconductor to shrink, increasing the short-circuit current of the module. However, the greatest effect that occurs in these conditions is the reduction in the open circuit voltage, causing the power, the form factor, and the performance also to reduce. Therefore, an increase in the operating temperature of the PV module reduces its efficiency, so it is necessary to extract as much energy as possible from the system to make it effective [30,29]. This leads to the need to develop strategies that counteract these adverse effects.

This type of problems is frequently found in various engineering branches, where it is required to establish the relationship between a given phenomenon and the parameters that influence it, in order to develop strategies that allow for decision making in operation processes, estimation of the phenomenon under certain conditions, and even the optimization of processes. The development of mathematical models provides a feasible solution for this type of situations. Mathematical models are a process of encoding and decoding of re-ality, in which a natural phenomenon is reduced to a formal numerical expression by a casual structure [24]. The modeling methods can be classified into statistical methods and heuristic methods (Fig. 10.1). The first are based on rigorous analysis of the interaction variables. They are the most widespread in the literature; however, they also turn out to be the most difficult to perform since a complete understanding of the interactions between the variables is required. On the other hand, heuristic methods are designed to establish an interaction between variables without a deep knowledge of the problem's physics, so they are commonly known as black-box models. The reason why the latter have experienced great growth in recent years is its quick implementation, making them one of the most popular of the artificial neural networks (ANNs).

The main disadvantage of the use of ANNs is their inability to realize an analysis of the importance of input variables on the model output [17]. Contrary to the statistical models, where the influence of each independent variable is expressed by the estimated coefficients of the model, the output of ANN models cannot be explained directly [28]. This is because in most occa-

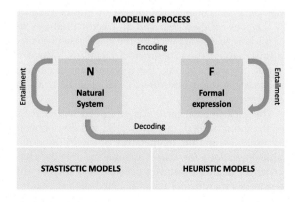

FIG. 10.1 Process of modeling and classification.

Artificial Neural Networks for Engineering Applications. https://doi.org/10.1016/B978-0-12-818247-5.00019-8

TABLE 10.1
Measured variables and instrumental equipment for the experimental database generation.

Measured variable		Sensor	Uncertainty
Global solar radiation	(G)	Campbell CS300-L Pyrometer	±5.00%
Environmental temperature	(T_a)	Campbell CS215-L	±4.00%
Wind velocity	(w_v)	WINDSONIC4-L	±2.00%
Wind direction	(w_d)	WINDSONIC4-L	±3.00%
Relative humidity	(RH)	Campbell CS215-L	±4.00%
PV voltage	(V_{PV})	TI LM747	±5.00%
PV current	(I_{PV})	FW BELL NT-50	±0.30%
PV temperature	(T_{PV})	Sensor K-Type	±0.80%

sions the hidden layer of ANNs is considered a black box. Nevertheless, in this scenario, the statistical techniques of sensitivity analysis are presented as an adequate complement for the study of artificial intelligence models such as ANNs, since among other things, they allow understanding of the nature of the internal representations generated by the network to respond to a given problem, demystifying the black-box concept.

In this chapter, the use of a global sensitivity analysis technique applied to multilayer neural networks is shown, using as a case study the prediction of temperature in the operation of a photovoltaic system. The tool proposed in this chapter represents a power complement for the artificial intelligence modeling since it provides a broader perspective in the interpretation of the results obtained in the so-called black-box models.

10.2 EXPERIMENTAL FACILITY AND DATABASE

A PV array installation comprised of monocrystalline PV Siemens SR100 panels and an electric dual-axis solar tracking system are used for the study case presented in this chapter. The experimental facility is located in Mérida, Mexico. This city has a warm humid climate; its latitude and longitude are 20° 58′ N and 89° 37′ W, respectively.

The experimental facility is equipped with sensors for the monitoring of meteorological variables, operating variables, and the temperature of the PV modules. Table 10.1 shows the sensed parameters and the characterists of the measuring devices used for the collection of experimental data. A meteorological station was placed in the vicinity of the experimental system to consider the direct effects of the environmental vari-

ables, i.e., environment temperature, wind direction, wind speed, and relative humidity. In addition, a pyranometer was mounted on the tracking structure to sense the overall global radiation that acts directly on the PV surface [22,14]. To measure the PV temperature, infrared temperature sensors are placed in contact with the PV panels' backside [12], where the mean of the PV array is used as the dependent variable. Finally, PV power data are indirectly calculated by the measured PV voltage and PV current.

The experimental database contains measurements made over a course of 3 years, collected every 30 minutes by a Campbell Scientific CR1000 datalogger. Fig. 10.2 shows a diagram with the experimental facility and the sample acquisition process.

10.3 SENSITIVITY ANALYSIS

Sensitivity analysis corresponds to a set of statistical techniques focused on determining how the variations of the M input variables ($x_i = x_1, x_2, \ldots, x_M$) of a mathematical model influence the response value (y_k). There are several reasons why sensitivity analysis is considered a necessary part in the development of computational models, specifically those based on artificial intelligence, including the following [9]:

- It allows to identify the order of importance that the input variables have for the output value generated by a mathematical model. This information is useful when one intends to use models based on ANNs as objective functions in optimization processes. This way it is possible to determine the feasibility of the optimization and to select in an appropriate way the operation variables that have the greatest impact on the process.

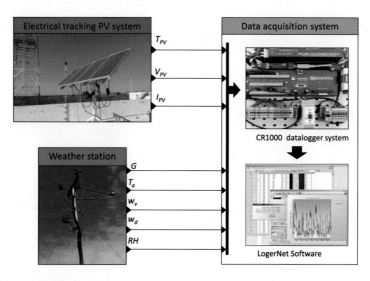

FIG. 10.2 Schematic diagram of the experimental facility designed to collect the case study database.

• It helps to identify variables that may need more research to improve the reliability of the model, because the ANNs are designed to establish correlations between dependent and independent variables without any notion of the nature of the data. By the analysis, it is possible to quickly evaluate various hypotheses of variables that are considered influential and determine their possible relevance, facilitating decision making in future studies (inverse process).

• It allows to identify the input variables that are insignificant for the model. In addition to knowing the range of importance of the input variables, the sensitivity analysis also identifies those that are not computationally relevant in the model. Therefore, these variables can be omitted, simplifying the computational model and reducing the computation time. This can be considered as an optimization process of the same mathematical model.

• It is helpful to determine if certain groups of input variables interact with each other. Some sensitivity analysis techniques allow us to identify how the input variables are affected by their mutual interaction. This interaction is useful at the moment of decision making to eliminate a variable from the computational model.

• It is useful when making decisions and identifying critical values. Since the ANN models are the product of the training and learning of experimental measurements, they are limited to operate between the maximum and minimum values of the input variables. Therefore, using sensitivity analysis it is possi-

ble to quantitatively know the acceptable limits for which the ANN model is functional.

• It helps to identify the coherence of the model with the real world and determine if the relationship is adequate. This is the most relevant point for ANN models, and artificial intelligence in general, because by its heuristic nature, sensitivity analysis serves as a tool to determine the adaptation of the model to the phenomenon under study, demystifying the black-box concept.

The above implies that at present the sensitivity analysis represents a vital element in the process of construction of ANN computational models. In order for a sensitivity analysis study to guarantee the correct reliability of its results, the method must meet the following specifications [24]:

• The method must be able to diagnose the effect of the input parameters, either individually or together.

• The method should test the influence of the input variables over the entire range.

• The method must be independent of the model.

• The method must be computationally efficient.

• The method must be applicable to discrete and continuous cases.

10.3.1 Sensitivity Analysis Classification

The literature reports various methods to carry out sensitivity analysis which can be grouped in different ways. However, one of the simplest ways to classify these methods is dividing them by the type of analysis performed and/or the type of interpretation of the results.

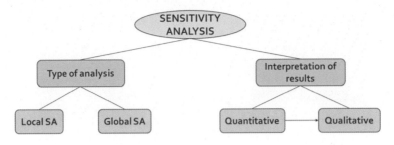

FIG. 10.3 Diagram of sensitivity analysis classification.

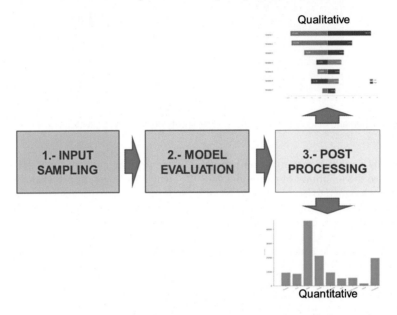

FIG. 10.4 Mandatory step to carry out a sensitivity analysis.

As shown in Fig. 10.3, when the classification is made based on the type of analysis, it can be subdivided into local sensitivity analysis (LSA) and global sensitivity analysis (GSA) [10,4]. The ASL implementation requires the modeler to specify a nominal or reference value $\bar{X}$ for the input variables. Thus, the variability of the model is given by the modification of each input value around the nominal value. On the other hand, GSA considers the variations of the entire possible ranges of values for the input variables [25,21]. Unlike LSA, GSA overcomes this possible limitation, but it still requires specifying the input variability space.

When the classification is conducted by the interpretation of results, it can be subdivided into quantitative and qualitative sensitivity analysis [26]. In qualitative sensitivity analysis, the sensitivity is assessed by visual inspection of model predictions or by visual tools (spe-cialized plots) such as tornado plots and scatter plots. On the contrary, quantitative sensitivity analysis focuses on presenting numerical values that represent the sensitivity index of the variables. Quantitative sensitivity analysis can be complemented with visual tools that help in its interpretation.

Regardless of the method used, all of them are governed by three essential steps for the computation of sensitivity. As indicated in Fig. 10.4, the first step consists in the generation of random samples for the input variables. In the second step, the random samples are evaluated in the computational model in order to obtain the response of the model to the perturbations. Finally, a postprocessing analysis is performed, where the results are analyzed quantitatively or qualitatively, depending on the case, in order to define the sensitivity of each variable.

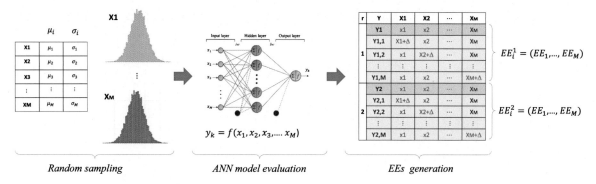

FIG. 10.5 Computational procedure for the generation of effective elements.

In this chapter we will focus on the study of one of the most versatile global sensitivity analysis techniques, known as the elementary effect test (EET). The following section aims to describe in a practical way this method of sensitivity analysis for its simple implementation in the environment of modeling engineering and artificial intelligence.

10.3.2 Elementary Effect Test

The EET is a GSA technique proposed by Morris [18] with low computation times, which makes it suitable for complex computational models such as ANNs. This technique allows ranking the importance of the input variables of a mathematical model in a simple way by using the so-called effective elements (EEs).

The EET works under the one-factor-at-a-time (OAT) concept, where the perturbations in the mathematical model are induced by varying one input factor at a time, keeping the other variables fixed. Fig. 10.5 illustrates the procedure for the generation of the EEs. In the first instance, a total of r random samples is generated for each input variable of the ANN model by applying numerical method techniques such as Monte Carlo or the Latin hypercube. The distribution of random samples is determined by the probability function used in sampling. The probability functions are chosen according to the nature of the input variable, where the most common are the uniform distribution function, the normal distribution function, and the Weibull distribution function. Subsequently, the random samples are grouped into data sets generating an $M \times r$ matrix of input variables. Each set of input variables ($x_i = x_1, x_2, \ldots, x_M$) is altered one element at a time by a defined perturbation value Δ. As can be seen in Fig. 10.5, by this alteration, for each set of input variables, M additional sets are generated. Therefore, the total number of ANN model

evaluations are given by

$$N = r(M + 1). \tag{10.1}$$

According to Fig. 10.5, once the N data set has been evaluated in the ANN model, the model responses to perturbations of the input variables are obtained. From these results it is possible to determine the EEs, considering the undistorted date sets as baseline points (X^j). In general it can be indicated that the EE of the ith input factor x_i at a given baseline point X^j and for a predefined perturbation is given by [13]

$$EE_i^j = \frac{y(x_1^j, x_2^j, \ldots, x_i^j + \Delta, \ldots x_M^j) - y(x_1^j, x_2^j, \ldots, x_i^j, \ldots x_M^j)}{\Delta}. \tag{10.2}$$

Thus, the EEs quantify how much the ANN result is affected by the disturbance in the ith variable, where the total number of EEs is r per variable. The EEs' average value (μ_i) is considered the sensitivity measure for the ith input parameter. In order to avoid compensation by negative values, the mean of the absolute values of EEs (μ_i^*) is used [7], i.e.,

$$\mu_i^* = \frac{1}{r} \sum_{j=1}^{r} \left| EE_i^j \right|. \tag{10.3}$$

On the other hand, as a secondary product of this analysis the intensity of the interaction of the ith input parameter with respect to another input parameter can be interpreted from the standard deviation of EEs (σ_i), i.e.,

$$\sigma_i = \frac{1}{r} \sum_{j=1}^{r} \left(EE_i^j - \mu_i \right)^2. \tag{10.4}$$

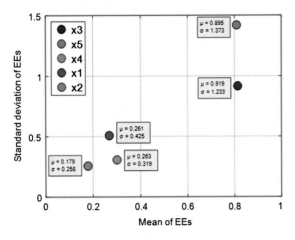

FIG. 10.6 Representation of an elementary effect test plot.

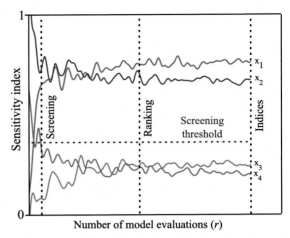

FIG. 10.7 Convergence analysis plot representing the three levels of convergence.

10.3.3 Elementary Effect Test Visualization

In order to establish a first immediate interpretation of the results by the sensitivity analysis method, the EET relies on two types of graphs, i.e., the EET plot and the convergence analysis plot, which provide different levels of information. An example of an EET plot is shown in Fig. 10.6. As can be seen, it is a bidimensional plot where the x-axis represents the mean absolute values of EEs (μ) and the y-axis represents the standard deviation of the EEs (σ) for each input variable. According to Fig. 10.6, on the x-axis, the greater the displacement of a variable towards the right, the greater the sensitivity or impact this variable will have on the output of the ANN model. Similarly, on the y-axis, the greater the upward displacement of a variable, the greater the influence it exerts on the other input variables. In the example of Fig. 10.6, it can be seen that a variable with a greater influence on the output of the model does not necessarily have the most influence among the input values, as seen with x_3 and x_5, as well as x_1 and x_4.

Fig. 10.7 illustrates an example of the convergence analysis plot. It estimates the sensitivity of each independent variable by systematically increasing the number of random samples (r) to be evaluated in the ANN model [20]. This type of visualization allows for verification that the sensitivity indices do not change, or vary slightly, when using different samples in the model evaluations, which are knowledge as convergence, guaranteeing the reliability of the results. In addition, it helps to determine the minimum number of r samples required for the sensitivity analysis of a given computational model.

The literature reports that for an adequate interpretation of the convergence analysis it is necessary to classify the convergence of the sensitivity analysis at three levels: convergence of screening, convergence of ranking, and convergence of sensitivity indices [26,21]. Fig. 10.7 exemplifies the points in which each of the convergence levels is met through the vertical lines screening, ranking, and indices:

- The first convergence level to be detected is the screening. At this level, the variables of the model are classified into influential parameters and noninfluential parameters. This classification is made using a screening threshold, where the variables below it are not considered relevant for the mathematical model. The screening threshold value is determined by the modeler; the most common values are 0.05 and 0.01. The convergence of screening is reached when the partition among sensitive and insensitive variables (indices below the screening threshold) is stabilized.
- The second convergence level is the ranking. The ranking is used to know the order of importance of the input variables in the ANN model. From this point, the analysis will focus on determining the value of the sensitivity index of each variable. This level of convergence is sufficient when the purpose is to determine only the importance of the variables, as in the case of optimization problems. The convergence of ranking is achieved when the ordering among the variables remains stable.
- The last convergence level is the determination of the sensitivity indices. At this level, the numeric value of the sensitivity index for each input variable is defined. However, it requires evaluating a large number of random samples to obtain them, making this the level of convergence most complicated to achieve.

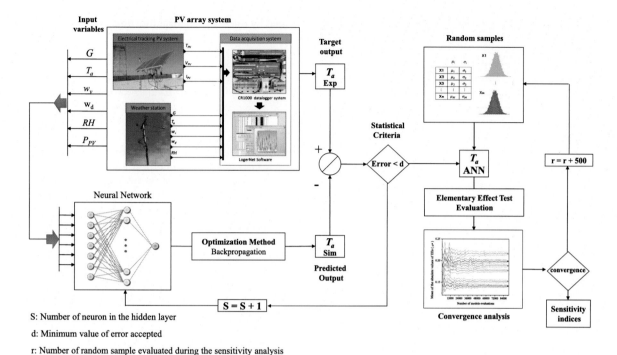

S: Number of neuron in the hidden layer

d: Minimum value of error accepted

r: Number of random sample evaluated during the sensitivity analysis

FIG. 10.8 Schematic workflow for the implementation of the EET with an ANN.

The more accurate the sensitivity index is desired, the greater the number of random samples (r) that will have to be evaluated in the model. The convergence of sensitivity indices is achieved when the values of the indices remain stable or quasistable.

10.4 APPLICATION

10.4.1 Modeling and Sensitivity Analysis Workflow

A computational workflow is shown in Fig. 10.8 indicating the steps to carry out ANN modeling together with EET. This workflow is based on the case study for predicting the operating temperature of PV modules by using multilayer perceptron neural networks. According to the figure, the workflow is divided into two parts: the first one corresponds to the development of the ANN model, and the second one focuses on the evaluation of the model by GSA methods to obtain the sensitivity indices.

In Fig. 10.8, the experimental database to develop the ANN model is randomly divided in two parts: 80% is destined to ANN learning and testing and the remaining 20% is used for the validation of the model.

This procedure guarantees an adequate distribution of the data and a good representation of the phenomenon under study. Once the database is divided, the learning process of the ANN is initiated. At this stage the database for training is normalized and entered into a basic ANN architecture, which is composed of M neurons in the input layer (the number of input variables), S neurons in the hidden layer (s), and k neurons in the output layer (the number of output variables). During the learning process, the ANN seeks to minimize the error between the estimated value and the experimental measurement by using optimization algorithms to adjust the weights and bias values, a technique commonly known as backpropagation [11]. When the learning process is completed, a statistical comparison between experimental data and ANN training/validation results is made. If it does not meet the criteria, the ANN architecture is modified more and the learning process is repeated. For the case study presented, Table 10.2 contains the ANN model parameter settings, such as the independent and dependent variables selected for the PV temperature modeling.

The sensitivity analysis is performed using the best ANN model obtained in the learning process. In the sensitivity analysis r random samples are generated by

TABLE 10.2
Parameter settings and variables characteristic for the development of artificial neural network PV prediction models.

ANN parameter	Setting
Hidden neurons active function	Tangent sigmoidal [2]
Output neurons active function	Pureline [2]
Optimization algorithm	Levenberg–Marquardt [8]
Maximum number of iterations	1000
Number of input neurons (M)	6
Number of output neurons (k)	1

Variable		Min	Max	Unit
Inputs:				
Global solar radiation	(G)	100	1367	[W/m^2]
Environment temperature	(T_a)	10	40	[°C]
Wind velocity	(w_v)	0.22	9.85	[m/s]
Wind direction	(w_d)	0	359	[°]
Relative humidity	(RH)	27	98	[%]
PV power	(P_{PV})	16.8	332	[W]
Output:				
PV temperature	(T_{PV})	12.7	66.4	[°C]

each input variable, based on a designed probability distribution function. Afterwards, the random samples are evaluated in the ANN model developed to generate the r effective elements. The next step is to compute the sensitivity indices for EE from 1 to r and to analyze the fulfillment of the three levels of convergence by a convergence analysis plot. If the comparison does not satisfy the convergence of sensitivity indices, the number of samples r to be evaluated is increased and the process is repeated iteratively.

10.4.2 Artificial Neural Network

The best ANN architecture for estimation of the temperature of the photovoltaic modules was obtained using a trial-and-error approach by varying the number of neurons (S) in the hidden layer, as described in Section 10.4.1. As can be seen in Fig. 10.9A, the developed ANN prediction model consists of 35 neurons in the hidden layer and the transfer function pair tansig–purelin, generating a 6-35-1 ANN architecture. It is de-

scribed by the mathematical function [6]

$$
\begin{aligned}
T_{PV,ANN} &= \sum_{s=1}^{S} \left[lw_{(1,s)} \left(\frac{1}{1 + \exp\left(-2\left(\sum_{m=1}^{M} \left(iw_{(s,m)}x_{(m)}\right) + b_{1(s)}\right)\right)} \right) + 1 \right] \\
&\quad + b_2,
\end{aligned}
\tag{10.5}
$$

where iw represents the weights from the input to the hidden layer, lw represents the weights from the hidden to the output layer, M is the total number of input neurons ($M = 6$), S is the total number of hidden neurons ($S = 35$), and b_1 and b_2 correspond to the bias values for the hidden and output neurons, respectively. This model was evaluated under the statistical parameters mean absolute percentage error (MAPE), root mean square error (RMSE), and coefficient of correlation (R2) [16]. The obtained results (MAPE = 3.48% and RMSE = 0.4541; the best linear behavior fitting with $R^2 = 98.48\%$) indicate that the model has a good prediction capacity with an accuracy above 93% (Fig. 10.9B).

An adequate practice in modeling with ANNs is to validate the developed model comparing its results with experimental samples not included in the training and testing phases. In the present study case, this validation was carried out using data measurements corresponding to standard days at the experimental location. From Fig. 10.10, it can be seen that the developed model has a good adaptation to the changes in temperature recorded during the day.

10.4.3 Sensitivity Analysis Results

To perform the sensibility analysis, it is necessary to consider the computational design set-up for the sampling of the variables and the convergence criteria. In this example, for the generation of the input variables, the uniform distribution function was selected using the normalized values present in Table 10.2 as function parameters. In addition, the sampling strategy used to proportionally distribute the random sample was the Monte Carlo method. On the other hand, a screening threshold of 0.1 and a maximum number of evaluated samples r equal to 100000 were taken as sensitivity analysis convergence criteria. Table 10.3 summarizes the selected settings for the sensitivity analysis.

Applying the methodology described in the workflow of Fig. 10.9, the sensitivity indices for the several cases of r are calculated. Fig. 10.11 exemplifies the convergence analysis results for three different r values (16000, 60000, and 90000). Fig. 10.11A illustrates the case when 16000 samples are used. As can be seen in

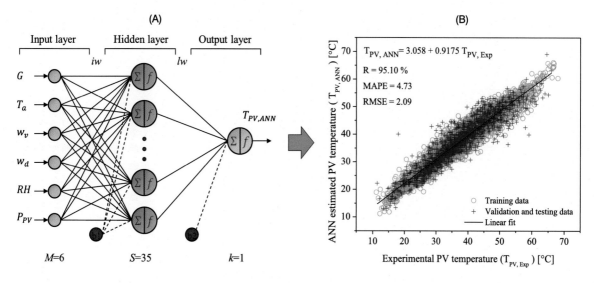

(A)   (B)

FIG. 10.9 Convergence analysis plot representing the three levels of convergence.

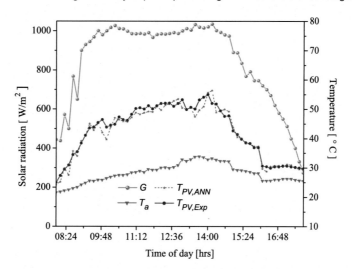

FIG. 10.10 ANN model validation with data not included in the training and testing phase.

TABLE 10.3
Sampling and convergence criteria set-up for the sensitivity analysis.

Sensitivity analysis set-up for sampling the parameter space	
Parameter distribution	Uniform distribution
Design type	Radial
Sample strategy	Monte Carlo
Sensitivity analysis set-up for convergence analysis	
Threshold value	1.0
Maximum number of samples evaluated (r_{max})	100000

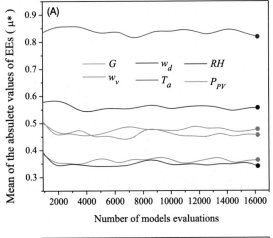

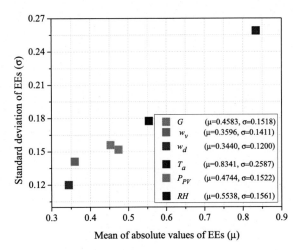

FIG. 10.12 Elementary effect test plots with the results of the case study.

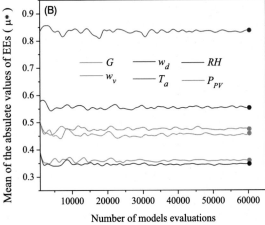

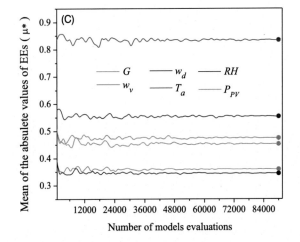

FIG. 10.11 Convergence analysis plots of the study case. **(A)** Convergence of screening achieved. **(B)** Convergence of ranking achieved. **(C)** Convergence of sensitivity indices achieved.

this figure, all the variables exceed the screening threshold assigned in Table 10.3; therefore it is considered that all are influential for the ANN model results. From this figure it is also distinguished that some parameters have to present a greater relevance in the model than others (T_a and RH). Nevertheless, according to the behavior of the convergence curves, the number of samples used does not indicate the general relevance of the variables. Fig. 10.11B shows the convergence analysis when $r = 60000$. As can be seen, the increase in the number of samples evaluated in the ANN model significantly improves the behavior of the convergence curves. At this point it is easy to see that the convergence of ranking has been fulfilled. However, the sensitivity indices have not been established accurately for some variables, such as RH and T_a. Finally, Fig. 10.11C presents the analysis for $r = 90000$. Similar to Fig. 10.11B, the increase in r reduces the disturbance in the sensitivity indices. From this point, it is observed that the disturbances in each sensitivity index are so insignificant that the values remain quasistable. Therefore, it can be concluded that this is the appropriate range for the case study that provides quantitative values of the sensitivity indices.

Once the convergence analysis with the three levels of convergence is satisfactorily concluded, it is possible to represent the results by an EET plot, as shown in Fig. 10.12. In this plot, it is seen that the ambient temperature is the parameter that has the greatest effect on the operating temperature of the PV module, and also the one that has the greatest influence on the other input variables. Also, from the analysis of the graph it is observed that although the solar radiation is less influential for the output of the model than the PV power, it has a greater impact on the input variables. Finally, the wind speed and direction are the parameters that contribute the least to the operating temperature of the module, according to the experimental measurements of the case study. Based on the case study, these results allow for identifying which variables affect the PV operation temperature to the greatest extent, and thus implementing the appropriate strategies to reduce their impact and improve their operation.

10.5 CONCLUSIONS

In this chapter, the implementation of a method of GSA in multilayer perceptron ANNs is presented. Using as a case study the modeling of the operating temperature of a PV module under the meteorological conditions of a city in the Yucatan peninsula, a methodology was proposed to link together the GSA and ANNs. The global sensitivity analysis technique presented in this chapter corresponds to the EET method, which due to its rapid response time, ranking capacity, and quantifiable value generation represents a feasible option to be implemented in the analysis of ANNs. In order to present a systematic implementation process, a computational workflow is proposed to facilitate the reader's understanding and use of the proposed methodology. The end of the chapter indicates that the use of the EET in ANNs allows the modeler to develop more complex models and facilitates the decision making in terms of operation, optimization, and analysis of the model's physics. Finally, the methodology presented in the workflow can be easily extrapolated for the use of more complex GSA techniques reported in the literature by using simple modifications.

REFERENCES

1. M. Bazilian, I. Onyeji, M. Liebreich, I. MacGill, J. Chase, J. Shah, D. Gielen, D. Arent, D. Landfear, S. Zhengrong, Re-considering the economics of photovoltaic power, Renewable Energy 53 (2013) 329–338.
2. M.H. Beale, M.T. Hagan, H.B. Demuth, Neural Network Toolbox™ User's Guide R2017a, The MathWorks, Inc., 2017.
3. E. Borgonovo, E. Plischke, Sensitivity analysis: a review of recent advances, European Journal of Operational Research 248 (3) (2016) 869–887.
4. E. Castillo, A.S. Hadi, A. Conejo, A. Fernández-Canteli, A general method for local sensitivity analysis with application to regression models and other optimization problems, Technometrics 46 (4) (2004) 430–444, http://www.tandfonline.com/doi/abs/10.1198/004017004000000509.
5. E. Cruz May, L.J. Ricalde, E.J.R. Atoche, A. Bassam, E.N. Sanchez, Forecast and energy management of a microgrid with renewable energy sources using artificial intelligence, in: C. Brito-Loeza, A. Espinosa-Romero (Eds.), Intell. Comput. Syst., Springer International Publishing, Cham, 2018, pp. 81–96.
6. Y. El Hamzaoui, M. Abatal, A. Bassam, F. Anguebes-Franseschi, O. Oubram, I. Castaneda Robles, O. May Tzuc, Artificial neural networks for modeling and optimization of phenol and nitrophenols adsorption onto natural activated carbon, Desalination and Water Treatment 58 (2017) 202–213.
7. L. Gao, B.A. Bryan, Incorporating deep uncertainty into the elementary effects method for robust global sensitivity analysis, Ecological Modelling 321 (2016) 1–9.
8. M.T. Hagan, M.B. Menhaj, Training feedforward networks with the Marquardt algorithm, IEEE Transactions on Neural Networks 5 (6) (1994) 989–993.
9. J.W. Hall, S.A. Boyce, Y. Wang, R.J. Dawson, S. Tarantola, A. Saltelli, Sensitivity analysis for hydraulic models, Journal of Hydraulic Engineering 135 (11) (2009) 959–969.
10. D.M. Hamby, A review of techniques for parameter sensitivity analysis of environmental models, Environmental Monitoring and Assessment 32 (2) (1994) 135–154.
11. S. Haykin, Neural Networks, Prentice Hall, Upper Saddle River, NJ, 1999.
12. M. Jankovec, M. Topič, Intercomparison of temperature sensors for outdoor monitoring of photovoltaic modules, Journal of Solar Energy Engineering 135 (3) (2013) 031012, http://solarenergyengineering.asmedigitalcollection.asme.org/article.aspx?doi=10.1115/1.4023518.
13. L. Lee, P. Srivastava, G. Petropoulos, Overview of sensitivity analysis methods in Earth observation modeling, in: Sensitivity Analysis in Earth Observation Modelling, 2017, pp. 3–24, http://linkinghub.elsevier.com/retrieve/pii/B978012803011000001X.
14. L.H.I. Lim, Z. Ye, D. Yang, Non-contact measurement of POA irradiance and cell temperature for PV systems, in: IECON 2015 - 41st Annu. Conf. IEEE Ind. Electron. Soc., 2015, pp. 386–391.
15. G. Makrides, B. Zinsser, M. Norton, G.E. Georghiou, M. Schubert, J.H. Werner, Potential of photovoltaic systems in countries with high solar irradiation, Renewable and Sustainable Energy Reviews 14 (2) (2010) 754–762.
16. O. May Tzuc, A. Bassam, M. Abatal, Y. El Hamzaoui, A. Tapia, Multivariate optimization of Pb(II) removal for clinoptilolite-rich tuffs using genetic programming: a computational approach, Chemometrics and Intelligent Laboratory Systems 177 (Jun 2018) 151–162, https://doi.org/10.1016/j.chemolab.2018.02.010, http://linkinghub.elsevier.com/retrieve/pii/S0169743917306044.
17. J.J. Montano, A. Palmer, Numeric sensitivity analysis applied to feedforward neural networks, Neural Computing & Applications 12 (2) (2003) 119–125, http://link.springer.com/10.1007/s00521-003-0377-9.

18. M.D. Morris, Factorial sampling plans for preliminary computational experiments, Technometrics 33 (2) (1991) 161–174.

19. J. Nossent, P. Elsen, W. Bauwens, Sobol' sensitivity analysis of a complex environmental model, Environmental Modelling & Software 26 (12) (2011) 1515–1525, http://www.sciencedirect.com/science/article/pii/S1364815211001939, http://www.sciencedirect.com/science/article/pii/S1364815211001939/pdfft?md5=a3e52e678ed8b2c2ca9cb2c613f039ea&pid=1-s2.0-S1364815211001939-main.pdf.

20. F. Pianosi, K. Beven, J. Freer, J.W. Hall, J. Rougier, D.B. Stephenson, T. Wagener, Sensitivity analysis of environmental models: a systematic review with practical workflow, Environmental Modelling & Software 79 (2016) 214–232, https://doi.org/10.1016/j.envsoft.2016.02.008.

21. F. Pianosi, T. Wagener, A simple and efficient method for global sensitivity analysis based on cumulative distribution functions, Environmental Modelling & Software 67 (2015) 1–11, https://doi.org/10.1016/j.envsoft.2015.01.004, http://www.sciencedirect.com/science/article/pii/S1364815215000237.

22. J. Polo, S. Garcia-Bouhaben, M.C. Alonso-García, A comparative study of the impact of horizontal-to-tilted solar irradiance conversion in modelling small PV array performance, Journal of Renewable and Sustainable Energy 8 (5) (2016).

23. M. Ratto, F. Campolongo, D. Gatelli, M. Saisana, Sensitivity analysis: from theory to practice, http://doi.wiley.com/10.1002/9780470725184.ch6, 2008, August 2015.

24. A. Saltelli, M. Ratto, T. Andres, F. Campolongo, J. Cariboni, D. Gatelli, M. Saisana, S. Tarantola, Global Sensitivity Analysis. The Primer, http://www.scopus.com/inward/record.url?eid=2-s2.0-84889461919&partnerID=40&md5=367cd0af66837fc6e4ae661b69007cd9, 2008.

25. A. Saltelli, P. Annoni, I. Azzini, F. Campolongo, M. Ratto, S. Tarantola, Variance based sensitivity analysis of model output. Design and estimator for the total sensitivity index, Computer Physics Communications 181 (2) (2010) 259–270, https://doi.org/10.1016/j.cpc.2009.09.018.

26. F. Sarrazin, F. Pianosi, T. Wagener, Global sensitivity analysis of environmental models: convergence and validation, Environmental Modelling & Software 79 (May 2016) 135–152, https://doi.org/10.1016/j.envsoft.2016.02.005.

27. W. Tian, A review of sensitivity analysis methods in building energy analysis, Renewable and Sustainable Energy Reviews 20 (Apr 2013) 411–419, https://www.sciencedirect.com/science/article/pii/S1364032112007101.

28. C. Vasilakos, K. Kalabokidis, J. Hatzopoulos, I. Matsinos, Identifying wildland fire ignition factors through sensitivity analysis of a neural network, Natural Hazards 50 (1) (2009) 125–143.

29. M.E. Ya'Acob, H. Hizam, T. Khatib, M.A.M. Radzi, C. Gomes, M. Bakri A., M.H. Marhaban, W. Elmenreich, Modelling of photovoltaic array temperature in a tropical site using generalized extreme value distribution, Journal of Renewable and Sustainable Energy 6 (3) (2014).

30. S. Zahurul, N. Mariun, M.L. Othman, H. Hizam, I. Zainal, A. Arash, Ambient temperature effect on Amorphous Silicon (A-Si) photovoltaic module using sensing technology, in: 2015 Ninth Int. Conf. Sens. Technol. Ambient, 2015, pp. 235–241.

CHAPTER 11

Pattern Classification and Its Applications to Control of Biomechatronic Systems

VICTOR H. BENITEZ, PHD

11.1 INTRODUCTION

Biomedical (BM) signals are generated by the activity of living beings and they are manifested as electrical or magnetic phenomena that can be recorded and analyzed. They constitute a rich source of data that encode important information about the living organ that generates such information.

Particularly in human beings, electrocardiography (ECG), electroencephalography (EEG), and electromyography (EMG) are examples of techniques developed to record the electrical activity of heart, brain, and muscles, respectively. The BM signals generated by ECG, EEG, and EMG techniques are mostly used to diagnose various diseases. However, in the last decade, an enormous amount of research has been done regarding the use of BM signals to develop human machine interfaces (HMIs) [18,11]. It is impossible to cover the vast number of applications related to BM signals. However, of the three techniques described above, EMG stands out for its applicability in the design of biomechatronic (BioM) systems.

BioM is concerned with the integration of biology disciplines into traditional mechatronics [24], i.e., BioM is related to the research and design of devices which integrate biology disciplines with mechanics, electronics, and control technology with assistive, diagnostic, or therapeutic purposes, or even to augment or to enhance human capabilities.

This chapter presents a methodology to process EMG signals in order to design a pattern classification module and to develop a BioM system capable of studying the finger movements of the human hand. The proposed approach is constituted by the following components:

- a neural network control module;
- a pattern classification module;
- a preprocessing module;
- an embedded system platform used as a testbed and a brushless DC motor as a servo mechanism.

The BioM system is designed to control brushless DC motors from myoelectric signals obtained from the forearm of healthy subjects. The EMG technique is employed to acquire and record the signals generated by five superficial muscles of the forearm; a preprocessing stage is carried out via Matlab and a control system is developed to generate trajectories applied to control brushless DC motors. The approach is tested both in simulation and in real-time via an embedded servo system developed in our laboratory. A key subject in the development of advanced BioM human machine interfaces based on EMG signals is the processing technique employed. EMG is an experimental technique concerned with the development, recording, and analysis of myoelectric signals. Myoelectric signals are formed by physiological variations in the state of muscle fiber membranes.

11.2 BIOMECHATRONIC SYSTEM COMPONENTS

The monitoring of muscle activity opens a vast panorama of lines of research to develop applications in various fields of engineering. Especially the development of control systems based on EMG signals has grown enormously in recent years [23,25,26]. A particular case of BioM systems is the so-called myoelectric control system, which uses exclusively EMG signals as inputs to design a control system. A myoelectric control system [3] is constituted by the following modules: a biological system, surface electrodes, signal conditioning, and signal processing.

11.2.1 Biological System

The electrical manifestation of the neuromuscular activation is directly related with muscle contraction. The myoelectric signals generated by muscle contraction can be detected by two types of electrodes: one type is located on the skin surface, the other is inserted as a nee-

Artificial Neural Networks for Engineering Applications. https://doi.org/10.1016/B978-0-12-818247-5.00020-4

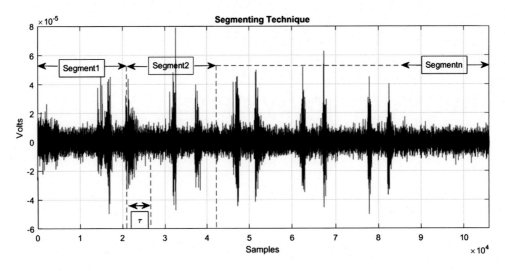

FIG. 11.1 Adjacent segmenting technique.

dle electrode. The signals recorded by means of the first technique are called superficial EMG (sEMG) signals. We use sEMG signals because this technique is less invasive for the user and the experiments can be done in a regular facility or laboratory without medical requirements.

11.2.2 Surface Electrodes

Electrodes consists of a metal detection surface that detects the electric current on the skin. Some electrodes include amplifier circuits with high input impedance; others are simple passive electrodes that adhere to the skin with specialized adhesive. The location of the electrodes plays a crucial role in EMG studies; considerations such as signal/noise ratio, signal reliability, and cross-talk must be taken into consideration to locate electrodes in the muscle of interest [14].

11.2.3 Signal Conditioning

This block is constituted by at least the following components:
- amplifier;
- filter;
- sampler.

The electrical manifestation of muscles goes from microvolts to millivolts; therefore, an amplifier stage is required to raise the magnitude of the signal to an adequate value.

11.2.4 Signal Processing

This block comprises three functional subprocesses:

- data segmentation;
- feature extraction;
- classification.

11.2.4.1 Data Segmentation

Studies performed by [10] highlighted the importance of segmenting the raw sEMG signal into time slots in order to extract features of myoelectric data. There are two popular techniques to segment raw sEMG data: adjacent windowing and overlapped windowing techniques [10]. In the first case, adjacent disjoint data segments with a predetermined size are used for feature extraction; with this technique, a processing delay is introduced after which each classified pattern emerges. In Fig. 11.1, the adjacent windowing technique is illustrated. The processing time delay is also depicted.

11.2.4.2 Feature Extraction

It is impractical to feed the classifier with the raw signal due to the large amount of data and the stochastic and nonstationarity nature of any biomedical signal. Instead, segmented data are mapped into a lower dimension feature vector whose entries are the features of the signal. According to the literature, features can be categorized in three groups: time frequency (TF), frequency, and time domain (TD) [27].

TD features are the ones most used in myoelectric classification due to their comparative computational simplicity, which in turn allows them to be implemented in real-time more easily. For the sake of completeness TD features are described as follows.

Mean of the Absolute Value

A calculation for mean absolute value (MAV) of the signal in segment i with N samples in length is given by

$$\overline{X_i} = \frac{1}{N} \sum_{k=1}^{N} |x_k|, \qquad (11.1)$$

where N is the total number of data points in a window segment, whose length is determined by the sampling frequency, and x_k is an EMG data value.

Root Mean Square

Root mean square (RMS) measures the signal energy, providing a strong measure of the information content of the signal. It is defined as

$$X_{RMS} = \sqrt{\frac{1}{N} \sum_{k=1}^{N} |x_k|^2}. \qquad (11.2)$$

Variance

Variance (VAR) is a representation of EMG signal power, it is related to the force developed by the muscle, and it is defined as

$$VAR = \frac{1}{N-1} \sum_{k=1}^{N} (x_k - \mu)^2. \qquad (11.3)$$

Zero Crossing

Zero crossing is the number of times that the waveform crosses zero; it is calculated for two consecutive samples given. Zero crossing increases if the following condition is fulfilled:

$$sign\left(-x_k \times x_{k+1}\right) and \left(|x_k - x_{k+1}| \geq threshold\right). \qquad (11.4)$$

Waveform Length

The waveform length gives information relative to the cumulative length of the waveform over the time segment and is defined as

$$WL = \sum_{k=1}^{N} |x_k - x_{k-1}|. \qquad (11.5)$$

Willison Amplitude

This feature represents, for every two consecutive data points, the number of times that the magnitude of the difference exceeds a threshold value. The Willison amplitude correlates to muscle contraction levels. It is defined as

$$W = \sum_{k=1}^{N} f\left(x_k - x_{k-1}\right), \qquad (11.6)$$

where

$$f(x) = \begin{cases} 1, & x > \quad threshold, \\ 0 & otherwise. \end{cases}$$

Energy

The area under the curve for two consecutive samples is determined as

$$E = \frac{1}{Fs} \sum_{k=1}^{N} |x_k|^2, \qquad (11.7)$$

where Fs is the sampling frequency.

Average Power

The average power (P) of a sequence with N samples is defined as

$$P = \frac{1}{N} \sum_{k=1}^{N} |x_k|^2. \qquad (11.8)$$

On the other hand, the TF representation of features allows to obtain a more accurate description of the physical phenomenon because TF analysis can localize the energy of the signal in both time and frequency. However, TF features are generally computationally expensive, which is a serious restriction in real-time applications. Therefore, linear, discrete TFs, such as short-time Fourier transform (STFT), wavelet transform (WT), and wavelet packet transform (WPT), are selected to extract features from EMG signals.

Short-Time Fourier Transform

Given a finite-length sequence x_k, $k \in \{0, ..., L-1\}$, with L as the length of the segment. Then its discrete Fourier transform (DFT) is defined as

$$X[mF] = \sum_{k=0}^{L-1} x[k] e^{-j2\pi(mF)(kTs)}, \qquad (11.9)$$

where $F = 1/LT_s$ is the frequency sampling step size. The STFT is obtained as a series of DTFs, indexed with respect to Ts, i.e.,

$$STFT[kT_s, mF] = \sum_{k=0}^{L-1} x[k] g[k-j] e^{-j2\pi mk/L}, \qquad (11.10)$$

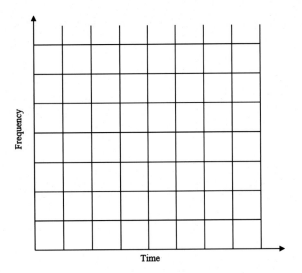

FIG. 11.2 Tiling of short-time Fourier transform.

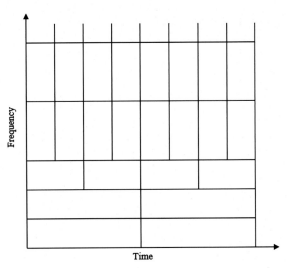

FIG. 11.3 Tiling of WT.

where $g[k]$ is the window function. The advantage of STFT is that it has a well-developed theory and can be computed very efficiently; the main drawback is that it requires each cell in the TF plane to have an identical shape, as depicted in Fig. 11.2.

Wavelet Transform

In the WT, by means of a variation in the TF ratio, it is possible to overcome the main drawback of the STFT. Additionally, with a good frequency resolution Δf in low frequencies and a good time localization Δt at high frequencies, it is possible to produce a tiling of the time–frequency plane that is appropriate for most BM signals (Fig. 11.3).

The continuous WT is defined as

$$CWT_x(\tau, a) = \frac{1}{\sqrt{a}} \int x(t) \Psi \left(\frac{t - \tau}{a} \right) dt, \qquad (11.11)$$

where $\Psi(t)$ represents the mother wavelet.

Wavelet Packet Transform

The WPT is a generalized version of the continuous WT and the discrete WT. The tiling of the time–frequency plane is configurable, as shown in Fig. 11.4.

A detailed review of the applicability of TF features is reported in [9].

11.2.4.3 Classification

It is well known in the literature that EMG signals are stochastic and nonlinear, with a nonstationary behavior; therefore, the properties of myoelectric signals

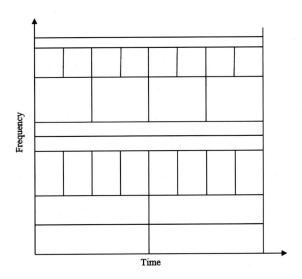

FIG. 11.4 Tiling of WT.

change over time. Due to the BM nature of myoelectric signals, it is reasonable to expect large variations in the values of a particular feature. Various external factors, such as variations in electrode location, muscular fatigue, muscle tone variation between individuals, gender of subjects, and sweat, cause changes in a signal pattern over time. The pattern classifier must be able to deal with such varying features optimally and overcome overfitting. If real-time requirements must be fulfilled, the classifier processing time should be fast enough to meet real-time constraints.

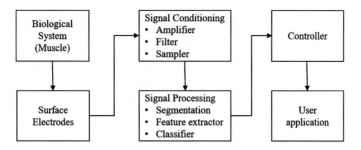

FIG. 11.5 Block diagram for myocontrol system.

11.2.5 Controller Deployment and User Application

Real-time constrictions and increasing demands for applications with high performance require a broad integration of technologies in the field of BioM such as powerful actuators, tiny gears, novel sensors, and miniature electronics. Advances in various fields of technology make it possible to integrate, for example, superflat BLDC motors with harmonic drivers and powerful 32-bit processors with smaller MOSFET power drivers. Prostheses, therapeutic and assistive/enhancing devices, requires BioM systems to fulfill requirements such as having a low weight and a small volume, being more human-like, having more flexible communication interfaces, having more manipulation capability, having more flexible stiffness in the joints, instead of rigid gears and linkage transmissions, allowing for high integration with the human body, containing a minimal amount of wires, and being capable of inertial sensing. Hardware architecture should meet high performance. Examples of powerful embedded systems are FPGAs, DSPs, and microcontrollers [16]. The building blocks of a myocontrol system are shown in Fig. 11.5.

11.2.6 Neural Network Control

Recurrent high-order neural networks (RHONNs) can be represented by

$$\dot{x}_i = -a_i x_i + \sum_{k=1}^{L_i'} w_{ik} z_{Iik} + \sum_{k=L_i'+1}^{L_i} w_{ij} z_{Iij}, \quad (11.12)$$

where $a_i > 0$, x_i is the state of the neuron, w_{ik} are the synaptic weights, and z_{Iik} are multiplicative terms of sigmoidal functions of x or u. On the other hand, w_{ij} are fixed weights, which allow us to incorporate a priori information about the structure of the mapping that we wish to model, z_{Iij} are functions of x or of u, L_i' is the number of synaptic weights, and L_i is the number of

neural connections of the ith neuron. The properties of (11.12) as identification and control scheme as well as its digital implementation have been widely described in works such as [5,4,2].

11.3 BIOMECHATRONIC SYSTEM PROPOSED

One of the main challenges in myoelectric control is to classify the signals issued by muscles, as they exhibit nonstationary behavior. Current approaches use myoelectric signals to characterize the position of the hand, where fingers are located in very well distinguishable positions. In [22], a study to decode individual flexion and extension movements of fingers is presented with 90% accuracy. They employed TD features to feed a multilayer perceptron used to classify 12 movements. Reference [15] presents a strategy to classify finger motion applied to movement recognition for Indonesian sign language systems, where a multimodal controller is developed to deal with uncertainties with individual variations in performing sign language. A naive Bayes classifier is implemented which reaches classification rates up to 98%. Another application for finger motion classifications can be found in [17], where a method for predicting individual finger movements from surface EMG is presented. They use just one TD feature, the mean average value, to develop a linear model of EMG feature dynamics, and Bayesian inference is used as classifier.

However, those approaches fail in characterizing small changes in finger movements. To the best of our knowledge, there are no reports in the literature on any methodology to study finger movements, whose angular motion varies with a small desired value. In the next sections, a method to distinguish several finger positions with very low mechanical fluctuations is presented. The methodology is verified with simulations and with a BioM system designed in our laboratory,

FIG. 11.6 The hand grasping an object in a scene.

which is controlled via superficial EMG signals generated by the forearm.

11.3.1 Problem Formulation

It is well known that in motion control, there are primitive variables such as position (θ), velocity (ω), acceleration (α), jerk (φ), and torque (τ). It is possible to establish such variables as dynamics of electromechanical servo systems [8]. These dynamics can be used to obtain complex motion patterns M, which can be represented as a set of nonlinear functions, i.e.,

$$M = f(\theta, \omega, \alpha, \varphi, \tau). \qquad (11.13)$$

Proposition. *We propose that it is possible to study finger motion by means of simple mechanical variables M in a controlled environment, i.e., a controlled sequence of continuous actions called scenes, as shown in Fig. 11.6.* □

In order to study the finger movement in a structured environment, let us introduce the term *scene*, which is defined here in below. The so-called scenes are constituted by movement sequences as follows.

Let n be a sequence of hand movements. A scene S is formally described as:

- Sequence 1: Hand performs a movement G_0, into an scene S with a period of time $t_0 < t \leq t_i$.
- Sequence k: Hand performs a movement $G_k = G_{k-1} + \Delta\Gamma$, into a scene S with a duration $t_i < t \leq t_k$, where $\Delta\Gamma$ is a structured variation in the Euclidian space (Fig. 11.6).
- Sequence n: Hand performs a movement $G_n = G_{n-1} + \Delta\Gamma$, into a scene S with a duration $t_{n-1} < t \leq t_n$.

Definition. A scene S is constituted by the set of movements

$$S = \{G_0, G_1, ..., G_k, ..., G_n\}. \qquad (11.14)$$
□

Considering that G_k terms are described by G_{k-1} plus a differential $\Delta\Gamma$, the movements are represented as follows:

$$\begin{aligned}
& G_0, \\
& G_1 = G_0 + k_1, \\
& G_2 = G_1 + k_2, \\
& G_3 = G_2 + k_3, \\
& \qquad \vdots \\
& G_n = G_{n-1} + k_{n-1}.
\end{aligned} \qquad (11.15)$$

Substituting backward and reordering, (11.15) is represented as

$$\begin{aligned}
G_0 &= initial\_cond, \\
G_1 &= G_0 + k_1, \\
G_2 &= G_0 + k_1 + k_2, \\
G_3 &= G_0 + k_1 + k_2 + k_3, \\
&\quad \vdots \\
G_n &= G_0 + K,
\end{aligned} \qquad (11.16)$$

where $K = k_1 + k_2 + ... + k_n$. Then, if (11.16) holds, the set S can be defined in terms of a triplet, i.e.,

$$S = \{G_0, G_n, K\}. \qquad (11.17)$$

In order to fulfill (11.17), the following definitions are made.

Definition. The metric of G_i is

$$\|G_i\| = G_{i-1} + k_i.$$
□

Definition. The differentials k_i are selected in such way that their values represent the anatomical variation of the subject being studied. □

Definition. All k_i terms are not necessary equals, but their sum is always equal to the final pose of fingers. □

As a consequence of the definition of K in (11.16), one can select the finger mechanical variation k_i as small as possible and then build up a model M as (11.13), according to the k_i selected values.

In the following sections, the definition of S, which constitutes a scene, is used to develop a structured environment such as grasping spheres with diameters D_i, which are selected with parameter variation k_i according to (11.16).

The range of movements (ROM) that the human hand can naturally perform is limited by several constraints, which can be grouped into three categories: static, intrafinger, and interfinger [7]. In this study only intrafinger constraints are considered because they are related to movement of flexion/extension of the fingers.

11.3.2 Experimental Preparation

Six healthy able-bodied subjects (four men, two women) aged 20–24 were trained to perform the experiment described in the next subsection. The experiment was conducted in accordance with the normative of the Ethics Committee of our institution; an ethical approval letter from the committee and consent of the volunteers was obtained prior to the experiment. The guidelines established in the surface EMG for the noninvasive assessment of muscles (SENIAM) project were followed (available at http://www.seniam.org).

Data were measured with a Trigno Wireless System (Delsys, Boston, Ma, USA); five channels were used in the experiment. Sampling frequency was set to 2000 Hz with a low-pass filter at 500 Hz; channel amplifiers were included in the experiment. Wireless RF electrodes were located on the arm, which was shaved, cleaned, and scrubbed with alcohol. Fig. 11.7 depicts the muscle electrode placement in the arm; the EMG system employed is illustrated in Fig. 11.8.

The designed experiment considered aspects such as determination of the type and amount of movements of the fingers, the number of participants, and the muscles of the forearm to be included, which will also influence the number of channels and dimensions of the characteristic vector as well as the quantity of objects to grasp by the hand.

11.3.3 Experimental Set-Up

Subjects were instructed to grasp five spheres randomly. The spheres were made of expanded polystyrene with diameters $D1 = 5.715$, $D2 = 6.985$, $D3 = 7.7978$, $D4 = 9.0678$, and $D5 = 10$ cm. Volunteers were seated comfortably and performed the requested grabbing of a sphere Di ($i = 1, ..., 5$). Electrodes were located on the following five superficial muscles: flexor carpi radialis,

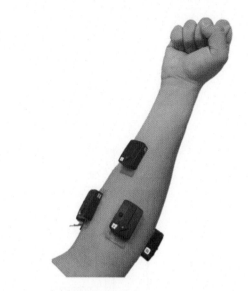

FIG. 11.7 Superficial muscles used to capture sEMG data.

FIG. 11.8 Hardware used to capture sEMG data.

flexor carpi ulnaris, flexor pollicis longus, flexor digitorum profundus, and extensor digitorum [14] (Fig. 11.7). The experiment was organized in batches; each subject performed a batch, which is established as the grasps of five spheres with a span of 5 seconds between each grasp. A relaxing period of 20 seconds between batches was introduced to avoid muscular fatigue. Subjects were prompted to perform six batch repetitions. Five repetitions were used as training data, one was used for validation. In order to synchronize the task of grasping and the recording of signals, subjects were trained to introduce a short contraction at the beginning and at the ending of grasping each sphere. The short contractions are used to get markers that can be easily handled by a maximum peak detector in the capturing system. The

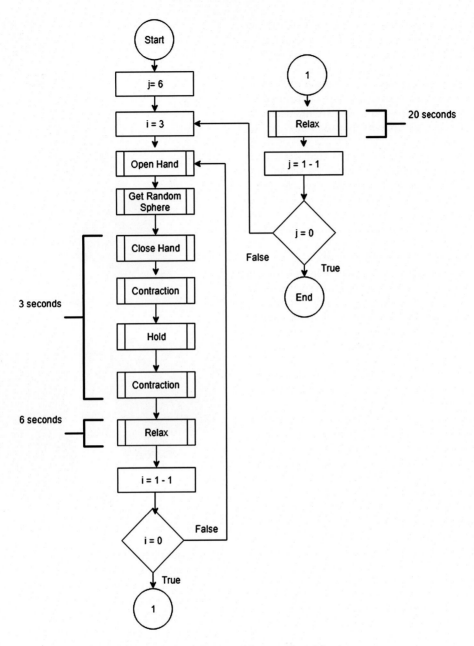

FIG. 11.9 Flow diagram of the experimental set-up.

whole batch can be summarized by the following tasks (Fig. 11.9):

- open hand,
- get sphere,
- close hand,
- make a short contraction,
- hold sphere i,
- make a short contraction,
- relax,
- repeat 1 to 6 three times,
- relax 20 seconds,
- repeat steps 1 to 8 thirty times.

TABLE 11.1
Joint angles for F1 (degrees).

Subject: F1
Weight: 56 kg
Age: 22 years
Height: 1.64 m

Sphere	Thumb		Index			Middle			Ring			Little		
	MP	PIP	MP	PIP	DIP	MP	PIP	DIP	MP	PIP	DIP	MP	PIP	DIP
D1	25	50	50	40	30	65	50	35	55	50	25	53	33	18
D2	24	45	40	40	25	55	50	30	50	40	25	43	20	20
D3	18	40	40	40	25	55	40	25	40	40	20	50	18	16
D4	15	40	35	30	15	40	40	25	40	28	20	33	15	15
D5	10	35	30	35	25	40	38	18	30	30	18	35	13	15

TABLE 11.2
Grasping sequence order for F1.

Experiment	Sequence of grasping
1	D1, D3, D4, D2, D5
2	D2, D5, D1, D3, D4
3	D4, D1, D3, D2, D5
4	D3, D1, D2, D5, D4
5	D5, D1, D3, D2, D4
6	D4, D3, D5, D1, D2

Before beginning the experiment, subjects were asked to grasp each sphere, and the angular positions of the fingers were measured with a goniometer. Four finger joints were measured for each sphere: the distal interphalangeal (DIP), proximal interphalangeal (PIP), metacarpophalangeal (MP), and trapezius-metacarpal (CMC) joints. The data for subject F1 are shown in Table 11.1. The grasping spheres $(D_1, .., D_5)$ order was determined randomly. Table 11.2 shows the grasping sequence order for F1.

11.3.4 Signal Processing
11.3.4.1 Data Segmentation
A segmentation process is performed in order to handle the data before feature extraction. Each sEMG signal is segmented considering a segment length of 256 ms [3]. The DC content of raw data is eliminated and normalized in order to improve the performance of the classifier because TD features are of different orders of magnitude. The min-max approach is selected to map

data to $\begin{bmatrix} 0 & 1 \end{bmatrix}$, which is expressed as

$$\bar{x} = \frac{x - \min x}{\max x - \min x}.$$

In Fig. 11.10, the raw data are shown for subject F1.

11.3.4.2 Feature Extraction
Instead of using raw signals, which is impractical due to the randomness of the sEMG signals, data are mapped into a lower dimension vector called a feature vector. Features are extracted from segmented data and used to feed the classification module. TD features are considered due to their computational simplicity, as demonstrated in the literature for myoelectric classification systems [3]. The TD features used are mean of the absolute value (MAV), root mean square (RMS), variance (V), zero crossing (ZC), waveform length (WL), Willison amplitude (WA), energy (E), and average power (AP), as described and detailed in [12,19,27]. Each feature is numbered with its corresponding muscle associated with a channel. Features were labeled as MAV1, MAV2,..., MAV5, RMS1, RMS2, and so on.

In Fig. 11.11, the RMS plot of normalized data for the flexor carpi ulnaris muscle is shown (subject F1). Notice the reduction of dimensionality obtained; the x-axis denotes time segments, where each segment represents a window of 255 ms.

Marker values are obtained as peaks in the feature vector and used to label the data according to Table 11.2. In Fig. 11.12 those markers are illustrated as peaks where D_i spheres were grasped.

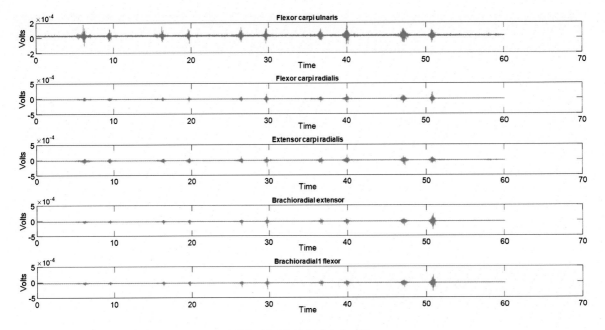

FIG. 11.10 Raw data for F1.

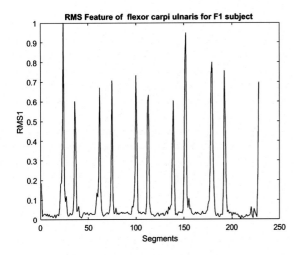

FIG. 11.11 Normalized RMS1 data for F1.

11.3.4.3 Classification
Four popular classifiers are considered: linear discrim-
inant analysis (LDA) [1], decision trees (DTs) [6], support vector machine with Gaussian kernel (SVM-GK),
and neural networks (NNs). Using raw sEMG signals
and eight TD statistics, a feature vector was determined for muscles and linking all together to arrange a
40-dimensional feature vector, which feed the classifiers.

Decision Trees
An accuracy up to 91% was reached with DT. However,
the classifier was unable to predict sphere 5 (B5). Predicted values are low for most of the spheres, except for
sphere 2. The false discovery rate is high. Fig. 11.13 displays the confusion matrix.

Linear Discriminant Analysis
Similar problems as with the DT occur with the LDA
classifier; some spheres are not predicted at all and the
false discovery rate is high, as can be seen in Fig. 11.14.

Support Vector Machine With Gaussian Kernel
This classifier's performance is the most convenient for
this application, as can be seen in Fig. 11.15. The confusion matrix shows high predicted rates, reaching an
accuracy of 97%.

11.3.5 Control of Biomechatronic System
11.3.5.1 Human Hand Biomechanics
The ROM that the human hand can naturally perform
is limited by several constraints, which can be grouped
into three categories: static, intrafinger, and interfinger
[7]. In this example only intrafinger constraints are considered because they are related to movement of flex-
ion/extension of the hand. The advances presented in
this study are restricted to the thumb tracking motion.

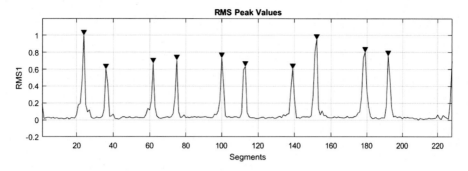

FIG. 11.12   RMS1 marker values for F1.

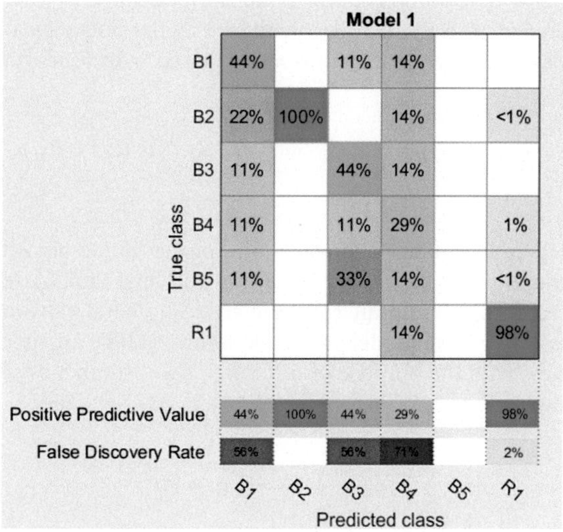

FIG. 11.13   DT confusion matrix.

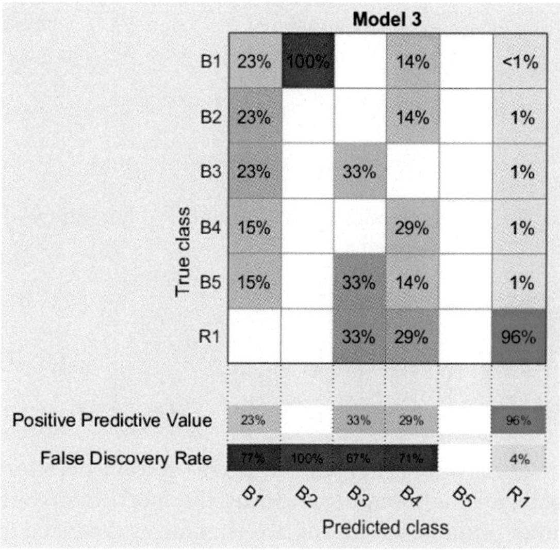

FIG. 11.14   LDA confusion matrix.

Thumb Biomechanics

The thumb is constituted by three bones: distal phalanx, proximal phalanx, and metacarpal, forming an open kinematic chain with three rotational joints: the interphalangeal (PIP), metacarpophalangeal (MP), and trapeziometacarpal (TMC) joints. This articulated system is shown in Fig. 11.16.

Without loss of generality, let us consider four positions to study the motion of the thumb (TP). Let us define the set

$$TP = \{OH, D_1, D_2, D_3\}, \qquad (11.18)$$

where OH and D_i for $i = 1, .., 3$ are the open hand, grasp sphere 1, grasp sphere 2, and grasp sphere 3 tasks, which restrict the hand at determined positions, as can be seen in Fig. 11.17.

Starting from OH, it is possible to design a motion path for the thumb angle positions while the hand grasps spheres of different diameters. Finger positions are defined in terms of their joint angles θ as defined in Table 11.1. Two joint angles θ for the thumb were chosen, so we have

$$\theta_{thumb} = \{\theta_{PIP}, \theta_{MP}\}. \qquad (11.19)$$

The positions defined in (11.18) introduce a dynamic in the kinematics of (11.19). A mapping that relates (11.18) and (11.19) is proposed as

$$\begin{aligned}
OH &= \{\theta_{PIP0}, \theta_{MP0}\}, \\
D_1 &= \{\theta_{PIP1}, \theta_{MP1}\}, \\
D_2 &= \{\theta_{PIP2}, \theta_{MP2}\}, \\
D_3 &= \{\theta_{PIP03}, \theta_{MP3}\}.
\end{aligned} \qquad (11.20)$$

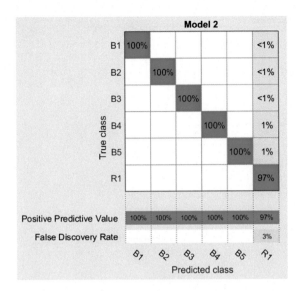

FIG. 11.15   SVM-GK confusion matrix.

The possible states of the thumb (TP*) are built with (11.20) and defined as

$$
TP^* = \begin{bmatrix} OH \\ D_1 \\ D_2 \\ D_3 \end{bmatrix} = \begin{bmatrix} \theta_{PIP0} & \theta_{MP0} \\ \theta_{PIP1} & \theta_{MP1} \\ \theta_{PIP2} & \theta_{MP2} \\ \theta_{PIP03} & \theta_{MP3} \end{bmatrix}. \quad (11.21)
$$

The rows of (11.21) can be used to design a trajectory path for the thumb considering the mechanical variables position, speed, and acceleration, as described in the problem formulation section for angular joints. In this case, a polynomial path that is able to track angular motion of TP* is designed and applied to track thumb motions via EMG signals.

11.3.5.2 Path Planning

Angular positions trained by the classifier are used to synthesize a desired reference. There are two methodologies to generate paths in robotics: polynomial and nonpolynomial methods. The polynomial approach is selected for its simplicity [13]. Let us consider a state machine with two discrete states $\theta_{PIP(k)}(t_0)$ and $\theta_{PIP(k+1)}(t_f)$, which are selected from (11.21). Then a parametric cubic polynomial path is defined as

$$
\theta_{IP} = a_0 + a_1 t + a_2 t^2 + a_3 t^3. \quad (11.22)
$$

Moreover, for any discrete point in (11.19), and considering (11.21), a polynomial function for two desired angular positions is represented as

$$
\theta_{thumb(k)}(t) = \begin{bmatrix} 1 & t_k & t_k^2 & t_k^3 \\ 0 & 1 & 2t_k & 3t_k^2 \\ 1 & t_{k+1} & t_{k+1}^2 & t_{k+1}^3 \\ 0 & 1 & 2t_{k+1} & 3t_{k+1}^2 \end{bmatrix}. \quad (11.23)
$$

11.3.5.3 Neural Controller

A multilayer feedforward neural network with time delay is trained to track the desired trajectories. The neural network can approximate a function whose parameters are the initial and final points of (11.21). The neural network is trained to learn trajectories defined by a reference model which is constituted by the polynomials described by (11.23) and parametrized with four constraints

$$
\theta_i(t_0) = \theta_{PIP0}, \quad \dot{\theta}_i(t_0) = \frac{d}{dt}\theta_{PIP0}, \quad \theta_i(t_f) = \theta_{MPf},
$$
$$
\dot{\theta}_i(t_f) = \frac{d}{dt}\theta_{MPf},
$$

where θ_i is defined for each joint of the thumb according to (11.21); t_0, t_f are the initial and final time of trajectories. The output y_n of the artificial neural network is applied to a model predictive control (MPC) architecture, with the purpose of minimizing a cost function J, defined as

$$
J = \sum_{i=N_1}^{N_2} \left[y_m(n+i) - y_n(n+i) \right]^2
$$
$$
+ \sum_{i=1}^{N_u} \alpha(k) [\Delta u(n+k)]^2, \quad (11.24)
$$

where N_1, N_2, and N_u are the costing horizons over which the error of the output and inputs are evaluated; y_m is the output of the reference model, and α is the contribution that the sum of the squares of the control has over J. The MPC approach is well known in the literature; the foundations were established in [21,20], where artificial neural networks were implemented as the plant nonlinear model.

The control scheme proposed is depicted in Fig. 11.18. A mismatch error function is added to guarantee that only the required positions are passed to the path planning module.

The mismatch function is expressed as

$$
M = \begin{cases} 0, & if \quad T = R, \\ 1, & if \quad T \neq R, \end{cases} \quad (11.25)
$$

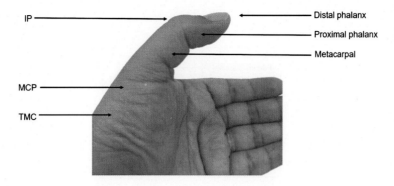

FIG. 11.16 Joints and phalanges of the thumb.

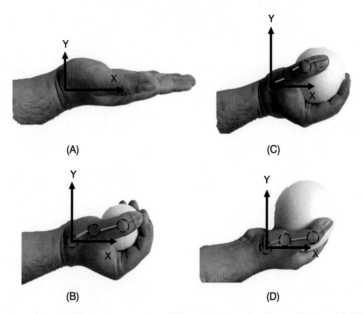

FIG. 11.17 Four motion positions for fingers. In **(A)** the hand is in the OH position. In **(B)–(D)** the hand is in D_1 to D_3, respectively. The x-axis is considered in the direction of the thumb when the palm is completely extended.

where T is vector generated by the classifier with correct prediction and R is the response vector generated with testing data. If $M = 1$, a biofeedback loop [3] is activated to indicate to the subject that a mismatch is found and it must repeat steps 1 to 7, as described by the experimental set-up.

11.3.5.4 Hardware Implementation

The hardware set-up is shown in Fig. 11.19. It consists of a Trigno Wireless System, as described in the Experimental preparation section. Data are processed by Matlab, where the classifier, previously trained, is selected

and the control algorithm is implemented. The output of the processing system is connected with a servo system, which integrates BLDC motors with a driver system with a microcontroller-based interface to receive control commands from Matlab.

11.3.5.5 Results

The SVM classifier is selected due to it demonstrated better performance. The BLDC motor is identified offline with a neural architecture constituted by 10 neurons in the hiding layer, two delayed plant inputs, and two delayed plant outputs with a sampling interval of

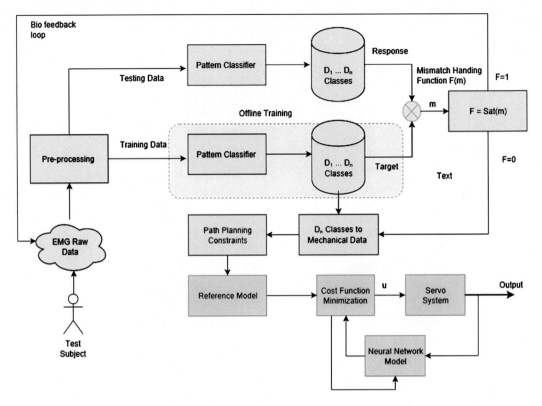

FIG. 11.18 Control scheme proposed.

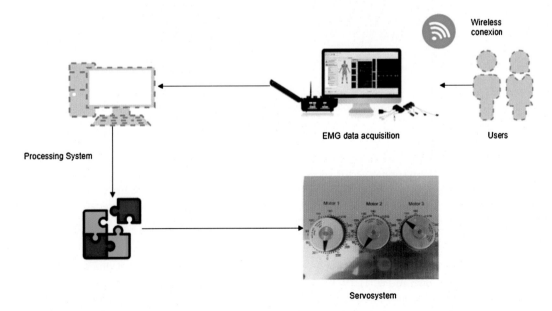

FIG. 11.19 Hardware implementation.

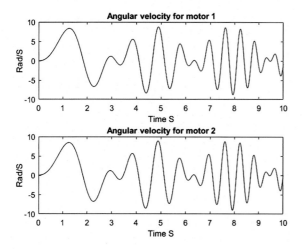

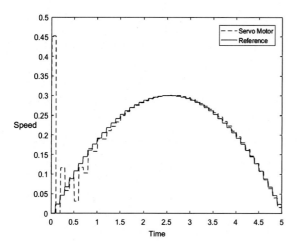

FIG. 11.20 Input/output pairs used to train the artificial neural network.

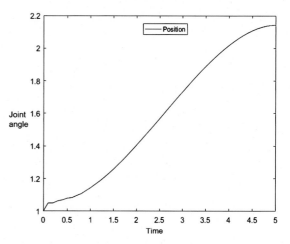

FIG. 11.22 Position of the MP joint.

11.4 CONCLUSION

The methodology proposed in this work can be applied to control electromechanical systems using superficial EMG signals. Moreover, the methodology proposed implements a new approach to study the motion of the fingers, which exhibits very small variations, which can be established by means of grasping spheres with diameters of predefined sizes. To the best of our knowledge, there is currently no methodology available that studies finger motions employing sEMG and considers small changes in the same movement.

FIG. 11.21 Tracking a velocity profile.

0.8 sec. Training data are constituted by 800 samples. Fig. 11.20 shows the input/output pairs. The MPC is configured with $\alpha = 0.005$, $N_2 = 8$, and $N_u = 2$; the minimization routine was based on the backtracking algorithm. Fig. 11.21 shows the velocity tracking profile applied to the MP joint (metacarpal link) of the thumb for subject F1 while performing a OH to D_i task. The speed profile corresponds to a difference of one degree, which is shown in Fig. 11.21, i.e., the metacarpal bone motion considered corresponds to a motion angle variation from 1 to 2.1 degrees. As can be seen in Fig. 11.22, the servo is able to track the position with good performance.

REFERENCES

1. A.H. Al-Timemy, G. Bugmann, J. Escudero, N. Outram, Classification of finger movements for the dexterous hand prosthesis control with surface electromyography, IEEE Journal of Biomedical and Health Informatics 17 (3) (2013) 608–618.
2. A.Y. Alanis, J.D. Rios, J. Rivera, N. Arana-Daniel, C. Lopez-Franco, Real-time discrete neural control applied to a linear induction motor, Neurocomputing 164 (2015) 240–251.
3. O.M. Asghari, H. Hu, Myoelectric control systems—a survey, Biomedical Signal Processing and Control 2 (4) (2007) 275–294.
4. V.H. Benitez, E.N. Sanchez, A.G. Loukianov, Neural identification and control for linear induction motors, Journal of Intelligent and Fuzzy Systems 16 (1) (2005).
5. V.H. Benitez, E.N. Sanchez, A.G. Loukianov, Decentralized adaptive recurrent neural control structure, Engineering Applications of Artificial Intelligence 20 (8) (2007) 1125–1132.
6. L. Breiman, Classification and Regression Trees, Chapman & Hall, 1993.

7. F. Chen Chen, S. Appendino, A. Battezzato, A. Favetto, M. Mousavi, F. Pescarmona, Constraint study for a hand exoskeleton: human hand kinematics and dynamics, Journal of Robotics (2013) 1–17.

8. J.N. Chiasson, Modeling and High Performance Control of Electric Machines, 1st ed., John Wiley & Sons, 2005.

9. K. Englehart, B. Hudgin, P.A. Parker, A wavelet-based continuous classification scheme for multifunction myoelectric control, IEEE Transactions on Biomedical Engineering 48 (3) (2001) 302–311.

10. K. Englehart, B. Hudgins, A robust, real-time control scheme for multifunction myoelectric control, IEEE Transactions on Biomedical Engineering 50 (7) (2003) 848–854.

11. S. Gupta, S.K. Chowdhary, Authentication through electrocardiogram signals based on emotions-a step towards atm security, in: 2017 6th International Conference on Reliability, Infocom Technologies and Optimization (Trends and Future Directions) (ICRITO), 2017, pp. 440–442.

12. B. Hudgins, P. Parker, R. Scott, A new strategy for multifunction myoelectric control, IEEE Transactions on Biomedical Engineering 40 (1) (1993) 82–94.

13. R.N. Jazar, Theory of Applied Robotics: Kinematics, Dynamics, and Control, Springer, 2010.

14. Z. Ju, H. Liu, Human hand motion analysis with multisensory information, IEEE/ASME Transactions on Mechatronics 19 (2) (2014) 456–466.

15. Khamid, A.D. Wibawa, S. Sumpeno, Gesture recognition for Indonesian sign language systems (ISLS) using multimodal sensor leap motion and Myo armband controllers based-on Naïve Bayes classifier, in: 2017 International Conference on Soft Computing, Intelligent System and Information Technology (ICSIIT), 2017, pp. 1–6.

16. H. Liu, K. Wu, P. Meusel, N. Seitz, G. Hirzinger, M.H. Jin, Y.W. Liu, S.W. Fan, T. Lan, Z.P. Chen, Multisensory five-finger dexterous hand: the DLR/HIT Hand II, in: 2008 IEEE/RSJ International Conference on Intelligent Robots and Systems, 2008, pp. 3692–3697.

17. N. Malesevic, D. Markovic, G. Kanitz, M. Controzzi, C. Cipriani, C. Antfolk, Decoding of individual finger movements from surface EMG signals using vector autoregressive hierarchical hidden Markov models (VARHHMM), in: 2017 International Conference on Rehabilitation Robotics (ICORR), 2017, pp. 1518–1523.

18. T. Mullen, C.A. Kothe, Y.M. Chi, A. Ojeda, T. Kerth, S. Makeig, T.-P. Jung, G. Cauwenberghs, Real-time neuroimaging and cognitive monitoring using wearable dry EEG, IEEE Transactions on Biomedical Engineering 62 (11) (2015) 2553–2567.

19. C. Sapsanis, G. Georgoulas, A. Tzes, D. Lymberopoulos, Improving EMG based classification of basic hand movements using EMD, in: 2013 35th Annual International Conference of the IEEE Engineering in Medicine and Biology Society (EMBC), 2013, pp. 5754–5757.

20. D. Soloway, P.J. Haley, Neural generalized predictive control, in: Proceedings of the 1996 IEEE International Symposium on Intelligent Control, 1996, pp. 277–282.

21. P.H. Sørensen, M. Nørgaard, O. Ravn, N.K. Poulsen, Implementation of neural network based non-linear predictive control, Neurocomputing 28 (1–3) (1999) 37–51.

22. F.V. Tenore, A. Ramos, A. Fahmy, S. Acharya, R. Etienne-Cummings, N.V. Thakor, Decoding of individuated finger movements using surface electromyography, IEEE Transactions on Biomedical Engineering 56 (5) (2009) 1427–1434.

23. L. Wang, S. Du, H. Liu, J. Yu, S. Cheng, P. Xie, A virtual rehabilitation system based on EEG-EMG feedback control, in: 2017 Chinese Automation Congress (CAC), 2017, pp. 4337–4340.

24. H. Witte, M. Fremerey, S. Weyrich, J. Mampel, L. Fischheiter, D. Voges, K. Zimmermann, C. Schilling, Biomechatronics is not just biomimetics, in: 9th International Workshop on Robot Motion and Control, 2013, pp. 74–79.

25. K. Xing, P. Yang, J. Huang, Y. Wang, Q. Zhu, A real-time EMG pattern recognition method for virtual myoelectric hand control, Neurocomputing 136 (2014) 345–355.

26. K. Xu, W. Guo, L. Hua, X. Sheng, X. Zhu, A prosthetic arm based on EMG pattern recognition, in: 2016 IEEE International Conference on Robotics and Biomimetics, ROBIO 2016, 2016.

27. M. Zecca, S. Micera, M. Carrozza, P. Dario, Control of multifunctional prosthetic hands by processing the electromyographic signal, Critical Reviews in Biomedical Engineering 30 (4–6) (2002) 459–485.

Index

Printed in the United States
By Bookmasters